G. Fichera (Ed.)

Autovalori e autosoluzioni

Lectures given at the
Centro Internazionale Matematico Estivo (C.I.M.E.),
held in Chieti, Italy,
August 1-9, 1962

C.I.M.E. Foundation
c/o Dipartimento di Matematica "U. Dini"
Viale Morgagni n. 67/a
50134 Firenze
Italy
cime@math.unifi.it

ISBN 978-3-642-10992-8   e-ISBN: 978-3-642-10994-2
DOI:10.1007/978-3-642-10994-2
Springer Heidelberg Dordrecht London New York

Printed on acid-free paper

Springer.com

CENTRO INTERNATIONALE MATEMATICO ESTIVO
(C.I.M.E)

Reprint of the 1ˢᵗ ed.- Chieti, Italy, August 1-9, 1962

AUTOVALORI E AUTOSOLUZIONI

ON EIGENVALUES EIGENFUNCTIONS AND RESOLVENTS

OF GENERAL ELLIPTIC PROBLEMS

Shmuel Agmon

Introduction

In these lectures we shall describe some recent results concerning the spectral theory of general non-self-adjoint elliptic boundary value problems. We shall be interested in the following problems: (i) Completeness of eigenfunctions. (ii) Angular distribution of eigenvalues. (iii) Asymptotic distribution of eigenvalues. The general plan of the lectures is as follows. In Lecture I we shall introduce the general class of regular elliptic boundary value problems and discuss the growth of certain resolvents in the complex plane. In Lecture II we shall establish completeness results for eigenfunctions of general elliptic problems obtaining also some results on the angular distribution of eigenvalues. In Lecture III we shall discuss some special classes of elliptic problems such as self-adjoint problems and absolutely elliptic problems. In Lecture IV we shall describe a very general result on the asymptotic distribution of eigenvalues of non-self-disjoint elliptic problems.

We note that the first three lectures are taken from the author's paper [1] which is due to appear shortly, whereas the material of the last lecture on the asymptotic distribution of eigenvalues is new.

S. Agmon

Lecture I

Regular Elliptic Boundary Value Problems and
Growth of Resolvents

We denote by G a bounded domain in n-space with boundary ∂G and closure $\overline{G}$. We let $x = (x_1, \ldots, x_n)$ be the generic point in E_n and use the notation:

$$D_i = \frac{\partial}{\partial x_i} \, , \qquad D = (D_1, \ldots, D_n) \, ,$$

denoting by

$$D^\alpha = D_1^{\alpha_1} \ldots D_n^{\alpha_n}$$

a general derivative. Here α stands for the multi-index $\alpha = (\alpha_1, \ldots, \alpha_n)$ whose length $\alpha_1 + \ldots + \alpha_n$ is denoted by $|\alpha|$.

We consider complex valued functions $u(x)$ defined in G (or $\overline{G}$). For $u \in C^j(\overline{G})$ we introduce the L_p norms ($p \geqslant 1$):

$$(1.1) \qquad \| u \|_{j, L_p(G)} = \left(\sum_{|\alpha| \leqslant j} \int_G | D^\alpha u |^p \, dx \right)^{1/p} \, .$$

The completion of $C^j(\overline{G})$ under the norm (1.1) is a Banach space of functions denoted here by $H_{j, L_p}(G)$. If the boundary is Lipschitzian, $H_{j, L_p}(G)$ coincides with the subclass of functions in $L_p(G)$ whose derivatives in the distribution sense of order $\leqslant j$ are functions belonging to $L_p(G)$.

We shall denote by $\mathcal{A}(x;D)$ an elliptic linear differential operator in $\overline{G}$ (variable complex coefficients) of even order 2m. Thus the characteristic

S. Agmon

polynomial associated with the principal part $\mathcal{A}'$ of $\mathcal{A}$ satisfies:

$$(1.2) \qquad\qquad \mathcal{A}'(x; \xi) \neq 0$$

for all real vectors $\xi = (\xi_1, \ldots, \xi_n) \neq 0$ and $x \in \bar{G}$. For n = 2 we shall also always assume that $\mathcal{A}$ satisfies the

ROOTS CONDITION. <u>For every pair of linearly independent real vectors</u> ξ, η <u>and</u> $x \in \bar{G}$ <u>the polynomial in</u> t: $\mathcal{A}'(x; \xi + t\eta)$ <u>has exactly</u> m <u>roots with positive imaginary parts</u>.

As is well known this condition is always satisfied if $n \geq 3$ or if n = 2 and the coefficients of $\mathcal{A}'$ are real.

We shall be interested in boundary value problems of the form:

$$\mathcal{A}(x; D)u(x) = f(x) \qquad \text{in } G \quad,$$

$$(1.3)$$

$$B_j(x; D)u(x) = 0 \qquad \text{on } \partial G, \quad j = 1, \ldots, m,$$

where $\left\{ B_j \right\}_{j=1}^m$ is a given system of m linear differential operators with coefficients defined on the boundary. We shall use the symbol $(\mathcal{A}, \left\{ B_j \right\}; G)$ to denote the boundary value problem (1.3) (omitting reference to the arbitrary given function f).

The general theory for higher order elliptic boundary value problems of the form (1.3) depends on suitable a priori estimates for the solution u. For these to hold it is necessary to restrict the class of problems by an algebraic condition. Denoting the principal part of B_j by B_j' this condition is the following:

COMPLEMENTING CONDITION. <u>At any point x of</u> ∂G <u>let</u> ν <u>denote the normal to</u> ∂G <u>and</u> $\xi \neq 0$ <u>a real vector parallel to the boundary.</u> <u>We require that the polynomials in</u> t, $B_j'(x; \xi + t\nu)$, $j = 1, \ldots, m$, <u>be linearly in-</u>

4

S. Agmon

dependent modulo the polynomial $\prod\limits_{k=1}^{m} (t - t_k^+(\xi)$ where $t_k^+(\xi)$ are the roots of $\mathcal{A}'(x; \xi + t\nu)$ with positive imaginary parts.

Suppose that the Complementing Condition holds, that the B_j are of order $m_j < 2m$, and that the domain and the differential operators satisfy the following

SMOOTHNESS ASSUMPTION. G is of class C^{2m}. The leading coefficients of $\mathcal{A}$ are continuous in $\bar{G}$, the other coefficients being measurable and bounded. The coefficients of B_j, $j = 1, \ldots, m$ belong to C^{2m-m_j} on the boundary.

Under the above assumptions the following a priori estimates hold:

THEOREM 1.1. Consider the class of functions u in $C^{2m}(\bar{G})$ satisfying the boundary conditions:

(1.4) $$B_j u = 0 \quad \text{on} \quad \partial G , \qquad j = 1, \ldots, m ,$$

and let $1 < p < \infty$. Then:

(1.5) $$\| u \|_{2m, L_p(G)} \leq C \left(\| \mathcal{A} u \|_{L_p(G)} + \| u \|_{L_p(G)} \right)$$

where C is some constant depending on $\mathcal{A}$, $\{ B_j \}$, G and p, but not on u.

A proof of this theorem in a more general situation is given in [5]. We shall denote by $H_{2m, L_p}(G; \{ B_j \})$ the completion in $H_{2m, L_p}(G)$ of the class of functions in $C^{2m}(\bar{G})$ satisfying the boundary conditions (1.4). Clearly Theorem 1.1 holds for all functions $u \in H_{2m, L_p}(G; \{ B_j \})$.

A boundary system of differential operators $\{ B_j \}$ is called a normal system if:

(i) The boundary ∂G is non-characteristic to B_j at each point.

(ii) The orders of the different operators are distinct.

S. Agmon

We introduce the following

DEFINITION 1.1.   <u>An elliptic boundary value problem</u> $(\mathcal{A}, \{B_j\}_1^m; G)$ <u>is called a regular problem if</u>

(i) <u>The elliptic operator</u> $\mathcal{A}$ <u>(of order</u> $2m$ <u>and satisfying the roots con-dition) together with the boundary system</u> $\{B_j\}$ <u>satisfy the Complementing Condition.</u>

(ii) $\{B_j\}_1^m$ <u>is a normal boundary system of</u> m <u>differential</u> <u>operators</u> of orders $\leq 2m-1$.

(iii) <u>The smoothness assumption on the domain and the coefficients</u> <u>introduced above holds.</u>

In the following all elliptic boundary value problems will be regular.

Let $(\mathcal{A}, \{B_j\}_1^m; G)$ be a regular elliptic problem and p some fixed number: $1 < p < \infty$. We shall denote by A the linear unbounded operator in $L_p(G)$ defined as follows:

(i) The domain of A is $\mathcal{D}_A = H_{2m, L_p}(G; \{B_j\}_1^m)$.
(ii) For u $\in \mathcal{D}_A$, Au $= \mathcal{A}(x;D)u$.

The operator A is clearly closed and it follows easily from the a prio-ri estimates (1.5) that the null space of A is finite dimensional and that its range is closed. If the spectrum of A is not the whole complex plane, i.e. if the resolvent:

$$(1.6) \qquad\qquad R(\lambda;A) = (\lambda I - A)^{-1}$$

exists for some $\lambda = \lambda_0$, then it follows readily (since $R(\lambda_0;A)$ is compact) that $R(\lambda;A)$ exists for all λ except for a discrete sequence of points $\{\lambda_n\}$ which are the eigenvalues of A. In general, however, one cannot exclude the possibility that the spectrum of A is the whole complex plane.

In the following we shall consider a subclass of regular problems for which it is possible to assert that the spectrum of A is discrete. In addition

S. Agmon

we shall obtain estimates for the growth of R(λ ;A) along certain rays in the complex plane. In this connection we introduce

DEFINITION 1.2 A ray arg $\lambda = \theta$ in the complex λ-plane is said to be a ray of minimal growth of R(λ ;A) if the resolvent exists for all λ sufficiently large on the ray, and if, moreover, for all such λ :

(1.7)
$$\| R(\lambda ;A)\| \leq \frac{c}{|\lambda|} ,$$

c > 0 a constant.

For regular elliptic boundary value problems $(\mathcal{A}, \{B_j\}_1^m ;G)$ one can determine the rays of minimal growth of the associated resolvent. The basic result here is the following

THEOREM 1.2. In order that the spectrum of A be discrete and the ray arg $\lambda = \theta$ be a ray of minimal growth of R (λ ;A) it is sufficient, and in case p = 2 also necessary , that the following two conditions be satisfied:

(i)
$$(-1)^m \frac{\mathcal{A}'(x; \xi)}{|\mathcal{A}'(x; \xi)|} \neq e^{i\theta}$$

for all real vectors $\xi \neq 0$ and all $x \in \bar{G}$.

(ii) At any point x of ∂G let ν be the normal vector and let $\xi \neq 0$ be any real vector parallel to the boundary at x. Denote by $t_k^+(\xi ; \lambda)$ the m roots with positive imaginary parts of the polynomial in t:

$$(-1)^m \mathcal{A}'(x; \xi + t \nu) - \lambda ,$$

where λ is any number on the ray arg $\lambda = \theta$. Then the polynomials (in t)

$$B_j'(x; \xi + t \nu), \qquad\qquad j = 1,\ldots,m ,$$

S. Agmon

are linearly independent modulo the polynomial

$$\prod_{k=1}^{m} (t - t_k^{+}(\xi ; \lambda)) .$$

In order to establish the sufficiency part of the theorem one needs to show that

(a) Under the conditions of Theorem 1.2 for all functions $u \in \mathcal{D}_A$ and all λ sufficiently large on the ray $\arg \lambda = \theta$, the following inequality holds:

$$(1.8) \qquad \| u \|_{L_p(G)} \leq \frac{\text{const.}}{|\lambda|} \, \| (A - \lambda) u \|_{L_p(G)} ,$$

(b) The range of $A - \lambda I$ is $L_p(G)$ for all λ sufficiently large on the ray.

Proof of (a): The inequality (1.8) is a special case of a more general result to be proved in [6] . We shall reduce the proof of (1.8) to a variant of Theorem 1.1 for a regular elliptic boundary value problem in $n+1$ variables. To this end introduce a new real variable t, put $D_t = \dfrac{\partial}{\partial t}$ and replace D by D_x. Consider the differential operator $\mathcal{L}$ in $n+1$ variables defined by:

$$(1.9) \qquad \mathcal{L}(x;D_x, D_t) = \mathcal{A}(x;D_x) - (-1)^m e^{i\theta} D_t^{2m} .$$

From condition (i) of Theorem 1.2 it follows that $\mathcal{L}$ is an elliptic operator of order $2m$ in the closure of the cylindrical domain

$\Gamma = \{(x,t) : x \in G, \ -\infty < t < \infty \}$. Moreover, it is readily checked that condition (ii) of the theorem is equivalent to the following: the elliptic operator $\mathcal{L}$ and the boundary system $\{B_j\}$ satisfy at each point of $\partial \Gamma$ the Complementing Condition introduced above. Consider the class of functions

S. Agmon

$v(x, t) \in C^{2m}(\bar{\Gamma})$ such that $v \equiv 0$ for $|t| \geq 1$, and

$$(1.10) \qquad B_j(x, D_x)v = 0 \quad \text{on} \quad \partial\Gamma \quad \text{for } j = 1, \ldots, m \ .$$

For functions v in this class the following a priori estimate holds:

$$(1.11) \qquad \|v\|_{2m, L_p(\Gamma)} \leq C\left(\|\mathcal{L}v\|_{L_p(\Gamma)} + \|v\|_{L_p(\Gamma)}\right),$$

where C is a constant. The estimate (1.11) follows from the localized version of Theorem 1.1. For a proof see proofs of Theorems 15.1 and 15.2 in [5] which implicitly contain this result (the corresponding result for the Schauder estimates is explicitly stated in [5] as Theorem 7.3).

Next, let $\zeta(t)$ be some fixed C^∞ function such that $\zeta(t) \equiv 0$ for $|t| \geq 1$, $\zeta(t) \equiv 1$ for $|t| \leq 1/2$. Given a function $u(x) \in C^{2m}(\bar{G})$ such that

$$(1.12) \qquad B_j u = 0 \quad \text{on} \quad \partial G , \qquad j = 1, \ldots, m ,$$

we define

$$(1.12)' \qquad v_\mu(x) = \zeta(t)e^{i\mu t} u(x) , \qquad \mu \text{ a real number} .$$

Denote by Γ_r the part of Γ in $|t| < r$. Since, clearly, the inequality (1.11) is applicable to v_μ, we have:

$$(1.13) \qquad \|v_\mu\|_{2m, L_p(\Gamma_1)} \leq C\left(\|\mathcal{L}v_\mu\|_{L_p(\Gamma_1)} + \|v_\mu\|_{L_p(\Gamma_1)}\right).$$

Now, $\mathcal{L}v_\mu = \zeta(t)e^{i\mu t}(\mathcal{A} - \mu^{2m}e^{i\theta})u$ + linear combination of derivatives of $u(x)e^{i\mu t}$ of order $\leq 2m-1$ with bounded coefficients. Using this (noting that $v_\mu \equiv ue^{i\mu t}$ for $|t| \leq 1/2$) we obtain readily from (1.13):

S. Agmon

$$\| ue^{i\mu t}\|_{2m,L_p(\Gamma_{1/2})} \leq \| v_\mu \|_{2m,L_p(\Gamma_1)}$$

(1.13)'

$$\leq C_1\left(\|(\mathcal{A}-\mu^{2m}e^{i\theta})u\|_{L_p(G)} + \sum_{j=0}^{2m-1} |\mu|^{2m-1-j}\|u\|_{j,L_p(G)}\right)$$

with a constant C_1 independent of μ or u. Also, we have:

$$\| ue^{i\mu t}\|^p_{2m,L_p(\Gamma_{1/2})} = \int_{\Gamma_{1/2}} \sum_{|\alpha|\leq 2m} |D^\alpha(ue^{i\mu t})|^p\, dxdt$$

(1.14)
$$\geq \int_G \sum_{k+|\beta|\leq 2m} |\mu|^{pk}|D_x^\beta u|^p\, dx$$

$$\geq \int_G \sum_{|\alpha|\leq j} |\mu|^{p(2m-j)}|D_x^\alpha u|^p\, dx =$$

$$= |\mu|^{p(2m-j)}\|u\|^p_{j,L_p(G)}$$

for any $j \leq 2m$. From (1.14) and (1.13)' we get:

$$\sum_{j=0}^{2m} |\mu|^{2m-j}\|u\|_{j,L_p(G)} \leq (2m+1)\|v_\mu\|_{2m,L_p(\Gamma_1)}$$

$$\leq (2m+1)C_1\|(\mathcal{A}-\mu^{2m}e^{i\theta})u\|_{L_p(G)}$$

S. Agmon

$$+ \frac{(2m+1)C_1}{|\mu|} \sum_{j=0}^{2m-1} |\mu|^{2m-j} \|u\|_{j,L_p(G)}$$

which gives

$$\left(1 - \frac{(2m+1)C_1}{|\mu|}\right) \sum_{j=0}^{2m} |\mu|^{2m-j} \|u\|_{j,L_p(G)}$$

(1.15)

$$\leq (2m+1)C_1 \|(\mathcal{A} - \mu^{2m}e^{i\vartheta})u\|_{L_p(G)}.$$

Putting $\lambda = \mu^{2m}e^{i\theta}$ it follows from (1.15) that for all $u(x) \in C^{2m}(\bar{G})$ satisfying the boundary conditions (1.12) and for all λ on the ray arg $\lambda = \theta$ such that $|\lambda|^{1/2m} = |\mu| \geq 2(2m+1)C_1$, the following inequality holds:

$$(1.16) \qquad \sum_{j=0}^{2m} |\lambda|^{\frac{2m-j}{2m}} \|u\|_{j,L_p(G)} \leq C_0 \|(\mathcal{A} - \lambda)u\|_{L_p(G)} ,$$

where $C_0 = 2(2m+1)C_1$ is a constant independent of u or λ. Clearly, by completion, (1.16) holds for all functions $u \in H_{2m,L_p}(G; \{B_j\})$ and in particular (1.8) follows. This establishes (a).

The estimate (1.8) shows in particular that for λ sufficiently large on the ray arg $\lambda = \theta$ the mapping:
$u \rightarrow (A - \lambda)u$ is a one-to-one mapping from $\mathcal{D}_A$ into $L_p(G)$. To complete the proof of the sufficienty part of the theorem one still has to show that the mapping is <u>onto</u>, or that the range of $A - \lambda I$ is the whole of $L_p(G)$. The proof of this fact is somewhat long and will be given in the author's paper on the existence theory for general elliptic boundary value problems [4]. We note that under additional smoothness assumptions one can also deduce this result from the existence results of Schechter [22] (in L_2) and Browder [8],

S. Agmon

and from the following observation: if $(\mathcal{A}, \{B_j\}; G)$ is sufficiently smooth then there exists a formally adjoint problem $(\mathcal{A}^*, \{B_j^*\}; G)$ which is regular and which, moreover, satisfies conditions (i) and (ii) of Theorem 1.2 with respect to the conjugate ray arg $\lambda = -\theta$.

Finally, let p = 2. We shall sketch the proof of the necessity part of the theorem. More precisely we shall show the following: if $(\mathcal{A}, \{B_j\}; G)$ is a regular problem such that the estimate (1.8) holds for all $u \in H_{2m, L_2}(G; \{B_j\}_1^m)$ then conditions (i) and (ii) of Theorem 1.2 must hold. To this end we first show that the regularity of $(\mathcal{A}, \{B_j\}; G)$ and (1.8) imply the stronger estimate (1.16). Indeed for $u \in H_{2m, L_2}(G; \{B_j\}_1^m)$ and λ sufficiently large on the ray arg $\lambda = \theta$ we have by assumption:

$$(1.17) \qquad \|u\|_{L_2(G)} \leq \frac{C_2}{|\lambda|} \|(\mathcal{A} - \lambda)u\|_{L_2(G)}$$

where here and in the following $C_2, C_3, \ldots$ denote constants not depending on λ. Also, applying the a priori estimate (1.5) and using (1.17), we get:

$$\|u\|_{2m, L_2(G)} \leq C_3 \left(\|\mathcal{A}u\|_{L_2(G)} + \|u\|_{L_2(G)} \right)$$

(1.18)

$$\leq C_3 \|(\mathcal{A} - \lambda)u\|_{L_2(G)} + C_3(|\lambda| + 1)\|u\|_{L_2(G)}$$

$$\leq C_4 \|(\mathcal{A} - \lambda)u\|_{L_2(G)}.$$

We shall use the following known inequality (see, for instance, Nirenberg [19])

S. Agmon

$$(1.19) \qquad \| u \|_{j, L_2(G)} \leq c \, \| u \|_{L_2(G)}^{1 - \frac{j}{2m}} \, \| u \|_{2m, L_2(G)}^{\frac{j}{2m}} \quad \text{for } j \leq 2m$$

where c is a constant depending only on G, n and m. Combining (1.19) with (1.17) and (1.18) we find that

$$(1.20) \qquad \| u \|_{j, L_2(G)} \leq C_5 \, |\lambda|^{\frac{j}{2m} - 1} \, \| (\mathcal{A} - \lambda) u \|_{L_2(G)}$$

for $j \leq 2m$ which is the same as (1.16).

Next consider the class of functions $v(x, t) \in C^{2m}(\overline{\Gamma})$ which are periodic in t, with period 2π, and which satisfy the boundary conditions: $B_j v = 0$ on $\partial\Gamma$, $j = 1, \ldots, m$. For such v let the Fourier series expansion in t be:

$$v(x, t) \sim \sum_{-\infty}^{\infty} u_n(x) e^{int} \ .$$

Clearly $u_n(x) \in C^{2m}(\overline{G})$, $B_j u = 0$ on ∂G for $j = 1, \ldots, m$, and we have the expansions:

$$D_x^\alpha D_t^k v \sim \sum_{-\infty}^{\infty} D_x^\alpha u_n(x)(in)^k e^{int} \ , \qquad |\alpha| + k \leq 2m \ ,$$

$$(1.21)$$

$$\mathcal{L}(x; D_x, D_t)v \sim \sum_{-\infty}^{\infty} \left(\mathcal{A}(x; D_x) - n^{2m} e^{i\theta} \right) u_n(x) \cdot e^{int} \ .$$

Also, by (1.20), we have:

13

S. Agmon

$$(1.22) \qquad \| u_n \|_{j, L_2(G)} \le C_5 |n|^{j-2m} \| (\mathcal{A} - n^{2m} e^{i\theta}) u_n \|_{L_2(G)}$$

for $|n| \ge N_0$. Applying now Parseval's formula to the derivatives of v, it follows readily from (1.22) and (1.21) and (1.5) that the class of functions v satisfy the following a priori estimate in Γ_π (the part of Γ in $|t| < \pi$):

$$(1.23) \qquad \| v \|_{2m, L_2(\Gamma_\pi)} \le \text{const.} \left(\| \mathcal{L} v \|_{L_2(\Gamma_\pi)} + \| v \|_{L_2(\Gamma_\pi)} \right).$$

A priori estimates of the type (1.23) in various norms were considered in [5] where it was shown that the validity of the a priori estimates implies that certain algebraic conditions must hold. By modifying somewhat the argument given in [5] one can show that the same result holds in our case and that the following conditions are necessary for (1.23) to hold: (i)' $\mathcal{L}$ is elliptic in (x, t) in $\bar{\Gamma}$. (ii)' $\mathcal{L}$ and the boundary system $\{ B_j \}$ satisfy the Complementing Condition on $\partial \Gamma$. Since conditions (i)' and (ii)' are equivalent to conditions (i) and (ii) of Theorem 1.2, the necessity part of the theorem follows.

S. Agmon

Lecture II

Completeness of Eigenfunctions and Angular
Distribution of Eigenvalues

Let $(\mathcal{A}, \{B_j\}_1^m ;G)$ be a regular elliptic boundary value problem of order 2m. For better clarity we shall denote by A_p the associated operator in $L_p(G)$ (previously denoted by A). Thus, $\mathcal{D}_{A_p} = H_{2m, L_p}(G; \{B_j\})$ and $A_p u = \mathcal{A}u$ for $u \in \mathcal{D}_{A_p}$. Suppose that the spectrum of A_p is not the whole plane so that, as was already remarked, it consists of a discrete sequence of eigenvalues $\{\lambda_k\}$. Let z_0 be some fixed point not in the spectrum of A_p. We put

$$(2.1) \qquad T_p = (A_p - z_0 I)^{-1} \quad .$$

Clearly T_p is a compact operator in $L_p(G)$ (it maps $L_p(G)$ into $H_{2m, L_p}(G)$) having the eigenvalues $\mu_k = 1/(\lambda_k - z_0)$. We also have the following relation between the resolvents of T_p and A_p :

$$(2.1)' \qquad R(\frac{1}{\lambda - z_0} ; T_p) = (\lambda - z_0)I - (\lambda - z_0)^2 R(\lambda ; A_p) , \qquad \lambda \neq \lambda_k .$$

An element $\phi \in L_p(G)$ ($\phi \neq 0$) is said to be a generalized eigenelement (eigenfunction) of T_p corresponding to the eigenvalue μ_k if $(T_p - \mu_k)^j \phi = 0$ for some integer $j \geq 1$. The smallest j for which the above relation holds is called the index of ϕ. As is well known the space of generalized eigenelements corresponding to an eigenvalue μ_k is finite dimensional and its dimension is called the multiplicity of μ_k.

15

S. Agmon

We shall denote by $sp(T_p)$ the closed subspace in $L_p(G)$ spanned by all generalized eigenelements of T_p.

The operator-valued function $R(\lambda; T_p)$ is a meromorphic function of $1/\lambda$ with poles at the eigenvalues μ_k. Let $f \in L_p(G)$ and consider the vector valued function $R(\lambda; T_p)f$ which is analytic in λ except for the origin and the points μ_k which are possibly poles. If $\lambda = \mu_k$ is a pole of $R(\lambda; T_p)f$ then in a sufficiently small neighborhood of μ_k one has the Laurent expansion:

$$(2.2) \qquad R(\lambda; T_p)f = \frac{\phi_1}{(\lambda - \mu_k)^j} + \frac{\phi_2}{(\lambda - \mu_k)^{j-1}} + \ldots + \frac{\phi_j}{\lambda - \mu_k}$$

$$+ \sum_{i=0}^{\infty} g_i (\lambda - \mu_k)^i ,$$

where $j \geq 1$, $\phi_1 \neq 0$ (the ϕ's and g's are elements in $L_p(G)$). Applying $T_p - \lambda$ to (3.2) one finds readily that: $(T_p - \mu_k) \phi_1 = 0$, $(T_p - \mu_k) \phi_2 = \phi_1, \ldots, (T_p - \mu_k) \phi_j = \phi_{j-1}$, so that ϕ_i is a generalized eigenelement of T_p of index i.

Similarly a function $\phi \in \mathcal{D}_{A_p}$ ($\phi \neq 0$) is said to be a generalized eigenelement (eigenfunction) of A_p corresponding to the eigenvalue λ_k if $(A_p - \lambda_k)^j \phi = 0$ for some $j \geq 1$ (one assumes, of course, that $\phi^{(1)} = (A_p - \lambda_k)\phi$, $\phi^{(2)} = (A - \lambda_k)\phi^{(1)}$, etc. belong to $\mathcal{D}_{A_p}$). The smallest integer j for which $(A_p - \lambda_k)^j \phi = 0$ is again called the index of ϕ. Clearly a function ϕ is a generalized eigenelement of A_p corresponding to the eigenvalue λ_k if and only if ϕ is a generalized eigenelement of T_p corresponding to the eigenvalue $\mu_k = \dfrac{1}{\lambda_k - z_0}$.

The closed subspace in $L_p(G)$ spanned by all generalized eigenelements of A_p is denoted by $sp(A_p)$. We shall say that the generalized eigenelements of the elliptic problem $(\mathcal{A}, \{B_j\}G)$ are complete in

S. Agmon

$L_p(G)$, if

$$(2.3) \qquad sp(A_p) = L_p(G) .$$

We shall first study the problem of completeness in $L_2(G)$. We shall need the following result on the growth of $R(\lambda;T_2)$ near the origin in terms of the dimension n of the underlying Euclidean space and the order $2m$ of the elliptic problem.

THEOREM 2.1. <u>Let</u> T_2 <u>be the above defined operator in</u> $L_2(G)$. <u>Then given</u> $\varepsilon > 0$ <u>there exists a sequence of positive numbers</u> $\rho_i \to 0$ $(i = 1, 2, \ldots)$ <u>such that</u> $R(\lambda;T_2)$ <u>exists everywhere on</u> $|\lambda| = \rho_i$ <u>and</u>

$$(2.4) \qquad \| R(\lambda;T_2) \| \le e^{\,|\lambda|^{-\frac{n}{2m}-\varepsilon}} \qquad \underline{for} \quad |\lambda| = \rho_i, \quad i = 1, 2, \ldots$$

The proof of this theorem is given in $[1]$ where it is deduced from a much more general result.

We now state the main completeness result in $L_2(G)$.

THEOREM 2.2. <u>Let</u> $(\mathcal{A}, \{B_j\}_1^m;G)$ <u>be a regular elliptic problem of order</u> $2m$. <u>Suppose that there exist rays</u> $\arg \lambda = \theta_j$, $j = 1, \ldots, N$, <u>in the complex plane such that</u>

(a) <u>The angles into which the complex plane is divided by these rays are all less than</u> $\dfrac{2m}{n}\pi$.

(b) <u>Conditions</u> (i) <u>and</u> (ii) <u>of Theorem</u> 1.2 <u>hold for</u> $\theta = \theta_j$, $j = 1, \ldots, N$.

<u>Then, the spectrum of the associated operator</u> A_2 <u>in</u> $L_2(G)$ <u>is discrete</u>. <u>Moreover, the generalized eigenfunctions of the elliptic problem are complete in</u> $L_2(G)$.

S. Agmon

<u>Remark.</u> We observe that if conditions (i) and (ii) of Theorem 1.2 hold for some $\theta = \theta_0$ then they also hold for all θ sufficiently near to θ_0. From this observation it follows that condition (a) of Theorem 2.2 could be replaced by the slightly weaker condition: (a)' The angles into which the complex plane is divided by the rays arg $\lambda = \theta_j$ are all $\leq \frac{2m}{n}\pi$. In particular if $m \geq n$ it suffices that conditions (i) and (ii) of Theorem 1.2 be satisfied for some number $\theta = \theta_0$ in order that the conclusion of Theorem 2.2 should hold.

Proof of Theorem 2.2 The discreteness of the spectrum of A_2 follows from Theorem 1.2. Moreover, the same theorem shows that the rays arg $\lambda = \theta_j$ are rays of minimal growth of $R(\lambda; A_2)$. That is, the resolvent exists on the rays for λ sufficiently large and

$$(2.5) \qquad \| R(\lambda;A_2) \| = 0(\frac{1}{|\lambda|}) \qquad \text{as } |\lambda| \longrightarrow \infty , \qquad \arg \lambda = \theta_j .$$

We shall show now that if $f^* \in L_2(G)$ is orthogonal to $\mathrm{sp}(A_2)$ then f^* is a null function. This will imply that $\mathrm{sp}(A_2) = L_2(G)$ or that the generalized eigenfunctions are complete in $L_2(G)$. To this end we may assume without loss of generality that the origin is not in the spectrum of A_2. Choosing in (2.1) $z_0 = 0$ we let: $T_2 = A^{-1}$. Consider now the function

$$(2.6) \qquad F(\lambda) = \overline{(f^*, R(\frac{1}{\lambda};T_2)f)}_{L_2(G)}$$

where f is some element in $L_2(G)$ (($\quad , \quad)_{L_2(G)}$ denoting scalar product in $L_2(G)$). From the properties of $R(\frac{1}{\lambda}; T_2)$ it follows that $F(\lambda)$ is an analytic function for $\lambda \neq \lambda_k$ where $\{\lambda_k\}$ is the sequence of eigenvalues of A_2. The points $\lambda = \lambda_k$ are either regular points or polar singularities of F. However, since f^* is orthogonal to all generalized eigen-

S. Agmon

elements of T_2 (we have $sp(T_2) = sp(A_2)$) it follows readily from (2.2) that the singular part in the Laurent expansion of $F(\lambda)$ around $\lambda = \lambda_k$ is zero. Thus F is also regular at the points $\lambda = \lambda_k$ and we conclude that $F(\lambda)$ is an entire function in the complex plane.

Next, from (2.6), (2.5) and (2.1)' (with $p = 2$, $z_o = 0$) we obtain the growth relations:

$$(2.7) \qquad |F(\lambda)| = 0(|\lambda|) \quad \text{as} \quad \lambda \longrightarrow \infty \text{ along the rays arg } \lambda = \theta_j,$$

$j = 1,\ldots,N$. Also, applying Theorem 2.1 it follows that for every $\varepsilon > 0$ there exists a sequence of positive numbers $r_i \longrightarrow \infty$, such that

$$(2.7)' \qquad |F(\lambda)| \leq e^{|\lambda|^{\frac{n}{2m} + \varepsilon}} \quad \text{for} \quad |\lambda| = r_i, \quad i = 1, 2, \ldots \ .$$

Consider now the function $F(\lambda)$ in the closure of any one of the angles into which the plane is divided by the rays arg $\lambda = \theta_j$, $j = 1,\ldots,N$. By assumption the size of the angle is $< \frac{2m}{n}\pi$. On the sides of the angle we have (2.7), and on a sequence of circles with radii tending to infinity the inequality (2.7)' holds. Choosing the number ε in (2.7)' sufficiently small we are in a position to apply the Phragmen-Lindelöf principle in the angle. It follows that in any such angle and consequently in the whole plane: $|F(\lambda)| = 0(|\lambda|)$ as $\lambda \longrightarrow \infty$. This in turn implies that F is a linear function: $F(\lambda) = c_o + c_1\lambda$. On the other hand it follows from (2.6) (using $R(\frac{1}{\lambda};T_2) = \lambda I + \lambda^2 T_2 + \ldots)$ that in the neighborhood of the origin:

$$F(\lambda) = \lambda \overline{(f^*,f)}_{L_2(G)} + \lambda^2 \overline{(f^*,T_2 f)}_{L_2(G)} + \ldots \ .$$

S. Agmon

This and the linearity of F give:

$$(2.8) \qquad\qquad (f^{*}, T_{2}f)_{L_{2}(G)} = 0 \; .$$

Since f is arbitrary while the range of T_2 is dense in $L_2(G)$, it follows from (2.8) that $f^{*} = 0$. This shows that $sp(A_2) = L_2(G)$ and establishes the theorem.

  <u>Example.</u> Let $\mathcal{A}$ be a second order elliptic operator in $\bar{G}$ with principal part:

$$\mathcal{A}' = -D_1^2 - D_2^2 - \ldots - D_{n-1}^2 - e^{i\gamma}D_n^2$$

where γ is a real number such that $0 \leq |\gamma| < \pi$. Consider the regular boundary value problem $(\mathcal{A}, B; G)$ where B is either the identity operator (Dirichlet problem) or $B = \dfrac{\partial}{\partial \ell} + a(x)$ where ℓ is a non-tangential (smoothly) variable direction and $a(x)$ is a smooth function on ∂G. One checks readily that in this case condition (b) of Theorem 2.2 holds for every ray $\arg \lambda = \theta$ which is outside the angle $0 \leq \arg \lambda \leq \gamma$ if $\gamma \geq 0$; the angle $\gamma \leq \arg \lambda \leq 0$ if $\gamma < 0$. Hence one can find a finite system of rays $\arg \lambda = \theta_j$ satisfying conditions (a) and (b) of Theorem 2.2 if $|\gamma| < \dfrac{2\pi}{n}$. Using the theorem we obtain the following result: <u>the generalized eigenfunctions of the above second order elliptic boundary value problem are complete in</u> $L_2(G)$ <u>if the number of variables</u> $n < \dfrac{2\pi}{|\gamma|}$.

  The completeness result of Theorem 2.2 implies in particular that under the same conditions the sequence of eigenvalues of the elliptic boundary value problem is infinite. A better result which gives information on the angular distribution of eigenvalues is the following

  THEOREM 2.3. <u>Let</u> $(\mathcal{A}, \{B_j\}_1^m; G)$ <u>be a regular elliptic problem</u>

S. Agmon

of order 2m. Suppose that the angle $\theta_1 < \arg \lambda < \theta_2$ in the complex plane $(0 < \theta_2 - \theta_1 \leq 2\pi)$ has the following properties:

    (a) Conditions (i) and (ii) of Theorem 1.2 hold for $\theta = \theta_i$, $i = 1, 2$.

    (b) $\theta_2 - \theta_1 < \dfrac{2m}{n}\pi$.

    (c) There exists a number θ_0 with $\theta_1 < \theta_0 < \theta_2$ such that either condition (i) or condition (ii) of Theorem 1.2 is violated for $\theta = \theta_0$.

Then the associated operator A_2 has infinitely many eigenvalues in the angle $\theta_1 < \arg \lambda < \theta_2$.

Proof: Using Theorem 1.2 it follows from (a) that the spectrum of A_2 is discrete. Also from the same theorem we get that the rays $\arg \lambda = \theta_i$, $i = 1, 2$, are rays of minimal growth of $R(\lambda; A_2)$ so that the resolvent exists on the rays for large λ and

$$(2.9) \quad \| R(\lambda; A_2) \| \leq \text{const.} \frac{1}{|\lambda|} \text{ for } \arg \lambda = \theta_i, \quad i = 1, 2, \quad |\lambda| \geq \Lambda_0 .$$

Next, let z_0 be some fixed point not in the spectrum of A_2 and put $T_2 = (A_2 - z_0 I)^{-1}$. Applying Theorem 2.1 to T_2 and using the relation $(2.1)'$ between the resolvents of A_2 and T_2 we conclude readily that given $\varepsilon > 0$ there exists a sequence of positive numbers $r_k \rightarrow \infty$, such that

$$(2.10) \quad \| R(\lambda; A_2) \| \leq e^{|\lambda|^{\frac{2}{2m}} + \varepsilon} \quad \text{for } |\lambda - z_0| = r_k, \quad k = 1, 2, \ldots$$

Suppose now by way of contradiction that the theorem is false. This would imply that there are only a finite number of eigenvalues of A_2 in the closed angle $\theta_1 \leq \arg \lambda \leq \theta_2$. Hence $R(\lambda; A_2)$ is analytic in the infinite sector $\sum : \theta_1 \leq \arg \lambda \leq \theta_2$, $|\lambda| \geq \Lambda_1$ if Λ_1 is chosen sufficiently large. Taking note of (2.10) and (b), choosing $\varepsilon > 0$ in (2.10) so small

21

S. Agmon

that $\theta_2 - \theta_1 < (\frac{n}{2m} + \varepsilon)^{-1}\pi$, we are in a position to apply the Phragmen–Lindelöf principle to the function $\lambda R(\lambda; A_2)$ in Σ. Since by (2.9) the function is bounded on the sides of Σ it follows that $\lambda R(\lambda; A_2)$ is bounded throughout Σ. In particular we get that

$$(2.11) \quad \| R(\lambda; A_2) \| \leq \text{const.} \ \frac{1}{|\lambda|} \ \text{for} \ \arg \lambda = \theta_0, \quad |\lambda| \geq \Lambda_1 \ .$$

But by the necessity part of Theorem 1.2 it follows from (2.11) that conditions (i) and (ii) of Theorem 1.2 hold for $\theta = \theta_0$. This, however, contradicts our assumption (c) and establishes the theorem.

We shall say that a ray $\arg \lambda = \theta_0$ is a direction of condensation eigenvalues of some given problem if for any $\varepsilon > 0$ the angle $|\arg \lambda - \theta_0| < \varepsilon$ contains infinitely many eigenvalues. From the last theorem we obtain the following

COROLLARY. Let $(\mathcal{A}; \{B_j\}, G)$ be a regular elliptic problem. Suppose that the ray $\arg \lambda = \theta_0$ satisfies the following conditions:

(a) Either conditon (i) or condition (ii) of Theorem 2.1 is violated for $\theta = \theta_0$.

(b) Conditions (i) and (ii) of Theorem 1.2 hold for all θ such that $|\theta - \theta_0| < \delta$, $\theta \neq \theta_0$, δ some positive number.

Then the ray $\arg \lambda = \theta_0$ is a direction of condensation of eigenvalues of A_2.

It is not difficult to extend the L_2 completeness result of Theorem 2.2 to general L_p and various other spaces as is shown in $[1]$. One gets the following more general result.

THEOREM 2.4. Under the conditions of Theorem 2.2 the generalized eigenfunctions of $(\mathcal{A}, \{B_j\}_1^m; G)$ are complete in $L_p(G)$ and are also

S. Agmon

<u>complete in</u> $H_{2m, L_p}(G; \{B_j\}_1^m)$ <u>in the</u> $\| \ \|_{2m, L_p}$ <u>norm for all</u>
$p < \infty$.

    <u>Remark.</u> Since the generalized eigenfunctions belong to
$H_{2m, L_p}(G; \{B_j\})$ for all p it follows from Sobolev's inclusion relations
(taking $p > n$) that the generalized eigenfunctions belong to $C^{2m-1}(\bar{G})$
and satisfy the boundary conditions $B_j \phi = 0$ on ∂G in the ordinary sense.
Using Sobolev's inequalities and Theorem 2.4 one can also show that any
function $f \in C^{2m-1}(\bar{G})$ satisfying the boundary conditions $B_j f = 0$
on ∂G $(j = 1, \ldots, m)$ can be approximated arbitrarily close in the
$C^{2m-1}(\bar{G})$ norm by finite sums of generalized eigenfunctions. Similar completeness results (under possibly additional smoothness assumptions on
$\mathcal{A}$, B_j and G) could be established in various other norms.

S. Agmon

Lecture III

Some Special Classes of Regular Elliptic Problems

We shall apply the previous results to some special classes of regular elliptic boundary value problems. We consider first the case of formally self-adjoint problems. A regular elliptic problem $(\mathcal{A}, \{B_j\}_1^m ;G)$ is said to be formally self-adjoint if the associated operator A_2 in $L_2(G)$ (with domain $\mathcal{D}_{A_2} = H_{2m, L_2}(G; \{B_j\}))$ is symmetric. From the symmetry it follows that for any $u \in \mathcal{D}_{A_2}$ and non-real λ:

$$(3.1) \qquad \|u\|_{L_2(G)} \leq \frac{1}{|\sin \theta| |\lambda|} \|(A_2 - \lambda)u\|_{L_2(G)}, \qquad \theta = \arg \lambda.$$

Let us recall that in the last part of the proof of Theorem 1.2 we have shown that inequality of the type (3.1) along some ray $\arg \lambda = \theta$ (with arbitrary constant) implies that conditions (i) and (ii) of Theorem 1.2 hold for θ . Hence we conclude that if $(\mathcal{A}, \{B_j\} ;G)$ is a regular formally self-adjoint elliptic problem then conditions (i) and (ii) of Theorem 1.2 hold for all θ such that $0 < |\theta| < \pi$. Using the sufficiency part of Theorem 1.2 it follows that the spectrum of the symmetric operator A_2 is discrete which in turn implies that A_2 is a self-adjoint operator.

Summing up we have established that if $(\mathcal{A}, \{B_j\}_1^m ;G)$ is a regular formally self-adjoint elliptic boundary value problem then the operator A_2 is a self-adjoint operator having a discrete real spectrum. We shall always assume in the following that $\mathcal{A}'$ (which is a real operator) is normalized so that

$$(3.2) \qquad\qquad (-1)^m \mathcal{A}'(x; \xi) > 0$$

S. Agmon

for all real $\xi \neq 0$ and $x \in \bar{G}$. Since now condition (i) of Theorem 1.2 is violated for $\theta = 0$ whereas conditions (i) and (ii) of Theorem 1.2 hold for all θ such that $0 < |\theta| < \pi$, it follows from the Corollary to Theorem 3.3 that the operator A_2 has infinitely many positive eigenvalues. We shall say the problem $(\mathcal{A}, \{B_j\} ; G)$ is bounded from below if the associated self-adjoint operator A_2 is bounded from below, or what amounts to the same thing if it has only a finite number of negative eigenvalues. Using again the Corollary to Theorem 2.3 and Theorem 1.2 one sees that this will be the case if and only if condition (ii) of Theorem 1.2 holds for $\theta = \pi$ (condition (i) obviously holds). This yields the following criterion for semi-boundedness:

THEOREM 3.1. <u>Let</u> $(\mathcal{A}, \{B_j\}_1^m ; G)$ <u>be a self-adjoint regular elliptic boundary value problem normalized by</u> (3.2). <u>Let</u> x^o <u>be an arbitrary point of</u> ∂G, ν <u>the normal vector at</u> x^o, <u>and</u> $\xi \neq 0$ <u>a real vector orthogonal to</u> ν . <u>Finally, for</u> $\lambda > 0$ <u>denote by</u> $t_k^+(\xi ; x^o, \lambda)$, $k = 1, \ldots, m$, <u>the</u> m <u>roots with positive imaginary parts of the polynomial in</u> t:

$$(-1)^m \, \mathcal{A}'(x^o; \xi + t\nu) + \lambda \, .$$

<u>With these notations, a necessary and sufficient condition for</u> $(\mathcal{A}, \{B_j\}_1^m ; G)$ <u>to be bounded from below is that the polynomials</u> (<u>in</u> t) $B_j'(x^o; \xi + t\nu)$, $j = 1, \ldots, m$, <u>be linearly independent modulo the polynomial</u>:

$$\prod_{k=1}^m (t - t_k^+(\xi ; x^o, \lambda))$$

<u>for all</u> $\lambda > 0$, $x^o \in \partial G$ <u>and</u> ξ <u>parallel to the boundary at</u> x^o.

We note that there exist regular self-adjoint elliptic boundary value problems which are not semi-bounded. This was shown in [3] where

S. Agmon

the general problem of semiboundedness was discussed from a different point of view.

Next consider a regular elliptic problem $(\mathcal{A}, \{B_j\}_1^m; G)$ which differs from a self-adjoint problem $(\tilde{\mathcal{A}}, \{\tilde{B}_j\}_1^m; G)$ only in lower order terms. That is the principal parts: $\mathcal{A}' = \tilde{\mathcal{A}}'$ and $B_j' = \tilde{B}_j'$. Since conditions (i) and (ii) of Theorem 1.2 depend only on the principal parts of the operators it follows from the properties of self-adjoint problems that $(\mathcal{A}, \{B_j\}_1^m; G)$ satisfies condition (i) of Theorem 1.2 for all θ such that $0 < |\theta| \leq \pi$ (one assumes (3.2)), and condition (ii) for θ such that $0 < |\theta| < \pi$. Applying Theorem 1.2, the Corollary to Theorem 2.3, Theorem 2.4 and Theorem 3.1 we obtain the following

THEOREM 3.2. <u>Let</u> $(\mathcal{A}, \{B_j\}_1^m; G)$ <u>be a regular elliptic pro-</u> <u>blem which differs from a (normalized) self-adjoint problem</u> $(\tilde{\mathcal{A}}, \{\tilde{B}_j\}_1^m; G)$ <u>only in lower order terms. Let</u> A_p $(1 < p < \infty)$ <u>be the</u> <u>unbounded operator in</u> $L_p(G)$ <u>associated with</u> $(\mathcal{A}, \{B_j\}_1^m; G)$. <u>Then:</u>

(i) <u>The spectrum of</u> A_p <u>is discrete, the eigenvalues and genera-</u> <u>lized eigenfunctions are common to all operators</u> A_p.

(ii) <u>All rays</u> $\arg \lambda = \theta$ <u>with</u> $0 < |\theta| < \pi$ <u>are rays of minimal</u> <u>growth of</u> $R(\lambda; A_p)$. <u>In particular there are only a finite number of eigen-</u> <u>values in any double angle:</u> $0 < \varepsilon \leq |\arg \lambda| \leq \pi - \varepsilon$.

(iii) <u>The positive axis is a direction of condensation of eigenvalues.</u>

(iv) <u>The negative axis is a ray of minimal growth of</u> $R(\lambda; A_p)$ <u>if</u> the condition of Theorem 3.1 <u>holds.</u> On the other hand if the condition does not hold then the negative axis is a direction of condensation of eigen- values.

(v) <u>The generalized eigenfunctions are complete in</u> $L_p(G)$; <u>they</u> <u>are also complete in</u> $H_{2m, Lp}(G; \{B_j\})$ <u>in the</u> $\| \ \|_{2m, Lp}$ <u>norm.</u>

We shall call a regular elliptic problem $(\mathcal{A}, \{B_j\}_1^m; G)$ absolute-

S. Agmon

<u>ly elliptic</u> boundary value problem if the boundary system $\left\{B_j\right\}_1^m$ has the property that the Complementing Condition of Lecture 1 is always satisfied no matter what is the elliptic operator $\mathcal{A}$ (subject to the "roots condition" if n = 2). An example of such a problem is the Dirichlet pro_ blem. More generally let

$$(3.3) \qquad B_j^{(s)} = \frac{\partial^{s+j-1}}{\partial\mu^{s+j-1}} + \text{ a lower order differential boundary operator,}$$

where s is some fixed integer, $0 \leq s \leq m$ and $j = 1, 2, \ldots, m$; $\frac{\partial}{\partial\mu}$ denoting differentiation at the boundary along the nontangential (smoothly variable) direction μ . Then, the "oblique derivative" problem $(\mathcal{A}, \left\{B_j^{(s)}\right\}_1^m ;G)$ is absolutely elliptic. Still another example of absolutely elliptic problem is when $B_j = \frac{\partial^{j-1}}{\partial\nu^{j-1}}$ for $j = 1, \ldots, m-1$ while B_m is a real normal boundary operator of order m. In the general case the algebraic structure of all absolutely elliptic boundary value problems was determined by Hörmander [17] who was also the first to introduce this class of elliptic problems.

For absolutely elliptic boundary value problems the statement of many of the results given previously becomes much simpler. The reason for this being that for such problems of condition (ii) of Theorem 1.2 is always a consequence of condition (i) and thus need not be stated. For instance, Theorem 1.2 for absolutely elliptic problems takes the following form.

THEOREM 3.3. <u>Let</u> $(\mathcal{A}, \left\{B_j\right\}_1^m ;G)$ <u>be an absolutely elliptic</u> <u>regular boundary value problem.</u> <u>Suppose that the following condition holds</u> <u>for some real number</u> θ :

S. Agmon

$$(3.4) \qquad (-1)^m \frac{\mathcal{A}'(x;\xi)}{|\mathcal{A}'(x;\xi)|} \neq e^{i\theta}$$

for all real $\xi \neq 0$ _and_ $x \in \bar{G}$. _Let_ A_p _be the associated unbounded operator in_ $L_p(G)$. _Then the spectrum of_ A_p _is discrete. Moreover, the ray_ $\arg \lambda = \theta$ _is a ray of minimal growth of_ $R(\lambda ;A_p)$. _For_ $p = 2$ _condition_ (3.4) _is also necessary for the above properties to hold._

Suppose that $(\mathcal{A}, \{B_j\}_1^m ;G)$ is absolutely elliptic and that in addition $\mathcal{A}'$ is a real operator normalized by (3.2). Then condition (3.4) holds for all θ such that $0 < |\theta| \leq \pi$. Hence, using the above form of Theorem 1.2, the Corollary to Theorem 2.3 and Theorem 2.4, we obtain the following results on eigenfunctions and eigenvalues of absolutely elliptic problems.

THEOREM 3.4. _Let_ $(\mathcal{A}, \{B_j\}_1^m ;G)$ _be an absolutely elliptic regular boundary value problem such that_ $\mathcal{A}'$ _is real and normalized by_ (3.2). _Let_ A_p _be the associated unbounded operator_ $L_p(G)$ $(1 < p < \infty)$. _Then:_

(i) _The spectrum of_ A_p _is discrete; the eigenvalues and generalized eigenfunctions are common to all operators_ A_p.

(ii) _All rays_ $\arg \lambda = \theta$ _different from the positive axis are rays of minimal growth of_ $R(\lambda ;A_p)$. _In particular there are only a finite number of eigenvalues outside any angle:_ $|\arg \lambda| < \varepsilon$, $\varepsilon > 0$.

(iii) _The positive axis is a direction of condensation of eigenvalues._

(iv) _The generalized eigenfunctions are complete in_ $L_p(G)$; _they are also complete in_ $H_{2m, Lp}(G; \{B_j\})$ _in the_ $\| \ \|_{2m, Lp}$ _norm._

Results on completeness of eigenfunctions and distribution of eigenvalues similar to ours were derived by Browder $[7 ; 9]$ in the case of the

S. Agmon

Dirichlet problem for higher order elliptic operators with a real principal part. The case of certain second order elliptic boundary value problems was considered earlier by Carleman [11] and Keldys [18] . We note however that all these special problems belong to the class of problems discussed in Theorem 3.2 and the proofs exploit the fact that the problems differ from a self-adjoint problem only in lower order terms. Theorem 3.4 on the other hand is applicable to non self-adjoint boundary value problems which are not obtained from a self-adjoint problem by a perturbation of lower order terms. A typical example of this kind is the oblique derivative boundary value problem for a real second order elliptic operator.

S. Agmon

Lecture IV

The asymptotic distribution of eigenvalues

In the preceding lectures we have discussed the completeness of eigenfunctions and the angular distribution of eigenvalues of general elliptic problems. In this lecture we shall discuss some results on the asymptotic distribution of eigenvalues.

Let $(\mathcal{A}, \{B_j\}; G)$ be a regular elliptic problem of order $2m$ and let A be the associated operator in $L_2(G)$ with domain $\mathcal{D}_A = H_{2m, L_2}(G; \{B_j\})$. Assume that the spectrum of A is discrete and let $\{\lambda_j\}$ be the sequence of eigenvalues of A each repeated according to its multiplicity. Without loss of generality we shall assume in the following that $\lambda = 0$ is not an eigenvalue and we shall put: $T = A^{-1}$. We first mention the following rather crude result concerning the distribution of the λ_j's but which holds for all regular problems. Namely, for every $\varepsilon > 0$, we have:

$$(4.1) \qquad \sum_j |\lambda_j|^{-\frac{n}{2m} - \varepsilon} < \infty .$$

To prove the theorem one notes that T is a compact operator in $L_2(G)$ with range contained in $H_{2m, L_2}(G)$ and with eigenvalues $1/\lambda_j$. By a general result proved in $[1]$ (Th. A1.1) the series (4.1) converges for any compact operator possessing the last mentioned property.

In order to obtain an <u>asymptotic</u> result on the distribution of eigenvalues it is necessary to restrict the class of regular problems considered. We shall not bother here to give results in the most general situation but shall limit ourselves to an important quite general subclass of problems.

S. Agmon

We denote by Z the subclass of regular problems such that the principal part of the elliptic operator is real and normalized by (3.2) and such that conditions (i) and (ii) of Theorem 1.2 hold for all directions different from the positive axis. For simplicity we shall also assume that the domain G is of class C^∞ and that all the coefficients of the differential operators are C^∞ functions. It follows from our previous results that if $(\mathcal{A}, \{B_j\}; G)$ belongs to Z then the associated operator A has a discrete spectrum with eigen-values λ_j condensing along the positive axis. It is possible to give an asymptotic formula for the number of eigenvalues with $\mathrm{Re}\,\lambda_j \leq t$. To this end, put:

$$N(t) = \sum_{\mathrm{Re}\,\lambda_j \leq t} 1 \quad ,$$

and define

$$w(x) = \int_{(-1)^m \mathcal{A}'(x;\xi) < 1} d\xi \qquad \text{and} \qquad w(G) = \int_G w(x)\, dx$$

we have :

THEOREM 4.1. Let $(\mathcal{A}, \{B_j\}; G)$ be an elliptic problem of order 2m belonging to the class Z. Then the following asymptotic formula holds.

$$(4.2) \qquad N(t) = (2\pi)^{-n} w(G) t^{\frac{n}{2m}} (1+o(1)), \quad t \to +\infty .$$

For second order self-adjoint elliptic boundary value problems the asymptotic formula (4.2) is classical (Weyl [23] and Courant [13]). Carleman [11] has introduced a powerful method which for second

S. Agmon

order problems yields the asymptotic formula also for certain non self-
adjoint problems. For self-adjoint problems associated with the biharmonic equation (plate problems) the asymptotic formula was established by
Courant [12] and Pleijel [21] . For higher order elliptic problems, but
in the special case of the self-adjoint Dirichelet problem, formula (4. 2)
was established by Gårding [15] (also Browder [10] for $2m > n$). The
method of Gårding [15] is applicable to certain other regular self-adjoint
and semi-bounded elliptic problems (see Ehrling [14]). Also, as was
pointed out by Garding, if (4. 2) holds for a regular self-adjoint and semi-
bounded problem, then (4. 2) also holds for the non-self-adjoint problem
obtained from the self-adjoint problem by a perturbation of lower order
terms. This follows from general results announced by Keldys [18] .

Our Theorem 4. 1. includes the above results as special cases.
It includes a wide class of non-self-adjoint problems which are not obtained from a self-adjoint problem by a perturbation of lower order terms.
Even for second order elliptic boundary value problems one has here a new
result in the case of the regular oblique derivative boundary value problem.

The proof of Theorem 4. 1 is base on the following result.

THEOREM 4. 2.   Under the conditions of Th. 4. 1 for every integer $k > n/2m$ the operator T^k is an integral operator given by a kernel $K(x, y)$ which is continuous on $G \times G$. Moreover, if $\Gamma (x, y; \lambda)$
is the resolvent integral operator of K (i. e. Γ is the kernel of
$T^k (I - \lambda T^k)^{-1}$), then $\Gamma (x, y; -t)$ exists for all positive t sufficiently large and

$$(4. 3) \qquad \int_G \Gamma (x, x; -t) \, dx = (2\pi)^{-n} w(G) \frac{\pi a}{\sin \pi a} t^{a-1} (1 + o(1)), \quad t \to +\infty$$

where $a = n/2mk < 1$.

S. Agmon

The deduction of Theorem 4.1 from Theorem 4.2 is easy. Indeed, if $D(\lambda)$ is the Fredholm determinant corresponding to the kernel K, then we have the well known relation:

$$(4.4) \qquad -\int_G \Gamma(x,x;\lambda)dx = \frac{D'(\lambda)}{D(\lambda)} \cdot$$

It is also well known that $D(\lambda)$ is an entire function of order ≤ 2 (actually ≤ 1) which in our case will have for zeros the k'th powers of the eigenvalues of A. Since by (4.1):

$$(4.5) \qquad \sum_j |\lambda_j|^{-k} < \infty$$

we have by Hadamard's factorization theorem of entire functions:

$$(4.6) \qquad \frac{D'(\lambda)}{D(\lambda)} = c_o + c_1 \lambda + \sum_j \frac{1}{\lambda - \lambda_j^k} \cdot$$

Since by (4.3) and (4.4): $D'(\lambda)/D(\lambda) \to 0$ for λ real $\to -\infty$, it follows readily that the constants c_o and c_1 in (4.6) are zero. Using this it follows from (4.6) and (4.3) that

$$(4.7) \qquad \sum_j \frac{1}{t + \lambda_j^k} = (2\pi)^{-n} w(G) \frac{\pi a}{\sin \pi a} t^{a-1}(1+o(1)), \quad t \to +\infty.$$

Let $\lambda_j = \mu_j + i\nu_j$. Since $\mu_j \to +\infty$ and $\nu_j / \mu_j \to o$, it follows from (4.7) and (4.5) that

S. Agmon

$$(4.8) \qquad \sum_{j} \frac{1}{t + \mu_j^k} = (2\pi)^{-n} w(G) \frac{\pi a}{\sin \pi a} t^{a-1}(1+o(1)), \quad t \to +\infty.$$

Applying now a Tauberian theorem of Hardy and Littlewood [16] to (4.8) one arrives at the asymptotic formula (4.2.).

Concerning the proof of Theorem 4.2 we note that from the re_ sults of Gårding [15] it already follows that for t=0 and for $t > 0$ sufficiently large there exists a kernel $\Gamma(x, y; -t)$ continuous on G x G, such that for all domains G_o with $\overline{G}_o \subset G$ and all functions having their support in G_o, one has:

$$T^k (I + t\, T^k)^{-1} f = \int_{G_o} \Gamma(x, y; -t)\, f(y) dy.$$

Also from the results of [15] it follows that for any such interior domain G_o one has:

$$(4.9) \qquad - \int_{G_o} \Gamma(x, x; -t) dx = (2\pi)^{-n} w(G_o) \frac{\pi a}{\sin \pi a} t^{a-1}(1+o(1)), \quad t \to +\infty.$$

To prove Th. 4.2 one uses (4.9) in conjunction with the following

THEOREM 4.2'. The kernel $\Gamma(x, y; -t)$ (for t=0 and all $t > 0$ sufficiently large) is continuous on $\overline{G}$ x $\overline{G}$ and one has the estimate:

$$\left(\int_{G} |\Gamma(x, x; -t)|^2 dx \right)^{1/2} \leq \text{Const. } t^{a-1}, \quad t \to +\infty.$$

Theorem 4.2', is this part in the proof of the asymptotic formula which takes into account the boundary conditions. In proving it we use

S. Agmon

Theorem 1.2 applied to the regular elliptic boundary value problem $(\mathcal{A}^k, \{B_j \mathcal{A}^i\} ; G)$ (j=1,...,m; i=0,...,k-1) as well as Sobolev type inequalities involving a parameter due to Ehrling [14]

S. Agmon

BIBLIOGRAPHY

[1] Agmon, S., On the eigenfunctions and on the eigenvalues of general elliptic boundary value problems, Comm. Pure Appl. Math., Vol. 15, 1962.

[2] Agmon, S., The angular distribution of eigenvalues of non self-adjoint elliptic boundary value problems of higher order, Conf. on Partial Differential Equations and Continuum Mechanics, The Univ. of Wisconsin Press, 1961, pp. 9-18.

[3] Agmon, S., Remarks on self-adjoint and semi-bounded elliptic boundary value problems, Proc. International Symposium on Linear Spaces, The Israel Academy of Sciences and Humanities, Jerusalem, 1961, pp. 1-13.

[4] Agmon, S., General elliptic boundary value problems, to appear.

[5] Agmon, S., Douglis, A., and Nirenberg, L., Estimates near the boundary for solutions of elliptic partial differential equations satisfying general boundary conditions. I, Comm. Pure Appl. Math., Vol. 12, 1959, pp. 623-727.

[6] Agmon, S., and Nirenberg, L., Properties of solutions of ordinary differential equations in Banach space, Comm. Pure Appl. Math., to appear 1963.

[7] Browder, F.E., On the eigenfunctions and eigenvalues of the general elliptic differential operator, Proc. Nat. Acad. Sci. U.S.A., Vol. 39, 1953, pp. 433-439.

[8] Browder, F.E., Estimates and existence theorems for elliptic boundary value problems, Proc. Nat. Acad. Sci. U.S.A., Vol. 45, 1959, pp. 365-372.

S. Agmon

[9] Browder,F.E., On the spectral theory of strongly elliptic differen_
tial operators, Proc. Nat. Acad. Sci. U.S.A. Vol. 45, 1959,
pp. 1423-1431.

[10] Browder, F.E., Le problème des vibrations pour un opérateur
aux derivées partielles self-adjoint et du type elliptique à coefficients variables, C.R. Acad. Sci. Paris 236 (1953), 2140-2142.

[11] Carleman, T., Über die Verleigung der Eigenwerte partieller
Differentialgleichungen, Ber. Verh. Sächś. Akad. Wiss. Leipzig.
Math. -Nat. Kl., Vol. 88, 1936, pp. 119-132.

[12] Courant, R., Über die Schwingungen eingespannter Platten,
Math. Zeitschrift, Vol. 15 (1922), pp. 195-200.

[13] Courant, R., and D. Hilbert, Methoden der Mathematischen Physïk
I, Berlin 1937.

[14] Ehrling, G., On a type of eigenvalue problems for certain elliptic
differential operators, Math. Scand. Vol. 2 (1954), pp. 267-285.

[15] Gårding, L., On the asymptotic distribution of the eigenvalues
and eigenfunctions of elliptic differential operators, Math. Scand.
Vol. 1 (1953) pp. 237-255.

[16] Hardy, G.H., and Littlewood, J.E., Notes on the theory of series
(XI): On Tauberian theorems, Proc. London Math. Soc. (2) Vol.
30 (1930), pp. 23-27.

[17] Hörmander, L., On the regularity of the solutions of boundary
problems, Acta Math., Vol. 99, 1958, pp. 225-264.

[18] Keldys, M.V., On the eigenvalues and eigenfunctions of certain
classes of non-self-adjoint equations, Doklady Akad. Nauk SSSR,
Vol, 77, 1951, pp. 11-14.

S. Agmon

[19] Nirenberg, L., _On elliptic partial differential equations_, Ann. Scuola Norm. Super. Pisa, Vol. 13, 1959, pp. 115-162.

[20] Å. Pleijel, _Propriétés asymptotiques des fonctions et valeurs propres de certains problèmes de vibrations_, Arkiv Mat. Astr. Fys. Vol. 27 A, No. 13 (1940), 100 pp.

[21] Å. Pleijel, _On the eigenvalues and eigenfunctions of elastic plates_, Comm. Pure Appl. Math. Vol. 3 (1950), pp. 1-10.

[22] Schechter, M., _General boundary value problems for elliptic partial differential equations_, Comm. Pure Appl. Math., Vol. 12, 1959, pp. 457-482.

[23] Weyl, H., _Über die asymptotische Verteilungsgesetz der Eigenwerte_, Göttinger Nachrichten, (1911), pp. 110-117.

CENTRO INTERNAZIONALE MATEMATICO ESTIVO

(C. I. M. E.)

A. M. OSTROWSKI

IL METODO DEL QUOZIENTE DI RAYLEIGH

ROMA - Istituto Matematico dell'Università

A. M. OSTROWSKI

IL METODO DEL QUOZIENTE DI RAYLEIGH[1]

Lezione I

Espressione approssimata delle autosoluzioni.

Sia Ω una matrice quadrata. Diremo che λ è un autovalore di Ω, se esiste un vettore ζ non nullo tale che $\Omega\zeta = \lambda\zeta$. Se λ è un autovalore, la matrice $\Omega - \lambda I$ è singolare.

Sia λ_0 un valore approssimato di λ; chiameremo autosoluzione approssimata relativa all'autovalore λ e all'approssimazione λ_0 la soluzione ζ_0 del sistema lineare

$$(1) \qquad \left(\Omega - \lambda_0 I\right)\zeta_0 = \eta, \qquad |\Omega - \lambda_0 I| \neq 0,$$

ove η è un vettore fisso scelto in modo generico (in un senso che verrà precisato).

Si può dimostrare che, quando $\lambda_0 \to \lambda$, $\zeta_0 / |\zeta_0|$ converge ad un autovettore relativo a λ. [2]

Sussiste infatti il seguente teorema più generale:

---

(1) Mi è sommamente grato esprimere la mia più viva gratitudine alla dott. Lucilla Bassotti e al prof. Luciano de Vito che hanno preso appunti delle mie lezioni e mi hanno prestato il loro esperto aiuto nella versione italiana di queste lezioni.

(2) Cfr. A. OSTROWSKI, Über näherungsweise Auflösung von Systemen homogener linearer Gleichungen, Zeitschrift für ang. Math. Phys. VIII, 1957.

A. M. Ostrowski

I. Sia S(t) una matrice dipendente analiticamente dal parametro t in un intorno J di t = o , tale che $|S(o)| = o$, $S(o) \neq o$ e per $t \in J - o$ riesca $|S(t)| \neq 0$; fissato comunque η fuori della varietà lineare descritta dal vettore $S(o)\xi$ al variare di ξ , e, detta $\zeta(t)$ la soluzione dell'equazione

$$(2) \qquad S(t)\,\zeta(t) = \eta ,$$

esiste una soluzione Π , diversa da zero, di $S(o)\,\Pi = o$ tale che

$$\lim_{t \to o} \zeta(t) \Big/ |\zeta(t)| = \Pi .$$

Dimostrazione

Il vettore $\zeta(t)$ è funzione analitica di t in J - o, dipendente razionalmente dagli elementi di S(t) ; per lo sviluppo in serie di LAURENT, si ha :

$$\zeta(t) = t^j \,(U + O(t)), \quad U \neq 0, \quad \text{(j finito ed intero)}$$

e quindi

$$|\zeta(t)| = |t|^j (|U| + O(t)) ;$$

donde

$$\frac{\zeta(t)}{|\zeta(t)|} = \left(\frac{t}{|t|}\right)^j \frac{U + O(t)}{|U| + O(t)} = \varepsilon \left(\frac{U}{|U|} + O(t)\right)$$

e pertanto

$$(3) \qquad \frac{\zeta(t)}{|\zeta(t)|} = \Pi + O(t),$$

A. M. Ostrowski

essendo Π un vettore unitario.

Se $j < 0$ si ha: $\lim\limits_{t \to 0} |\zeta(t)| = +\infty$; mpltiplicando la (3) a sinistra per $S(t)$, si ottiene :

$$(4) \qquad \frac{\eta}{|\zeta(t)|} = S(t)\,\Pi + S(t)\,O(t)$$

e quindi, per $t \to 0$, $S(o)\,\Pi = o$.

Se $j = o$, si ha $\lim\limits_{t \to 0} |\zeta(t)| = |U|$ e quindi passando al limite per $t \to 0$ nella (4) :

$$(5) \qquad \eta = S(o)\,\Pi\,|U| \ .$$

Se si sceglie η in modo che non sia della forma $S(o)\,\zeta$, la (5) non può essere verificata.

Se infine $j > o$, si ha $\lim\limits_{t \to 0} \zeta(t) = o$ e di conseguenza

$o = \lim\limits_{t \to 0} S(t)\,\zeta(t) = S(o)\,U = o \neq \eta$; ciò è contro l'ipotesi.

Osserviamo che, per l'ipotesi $S(o)$ degenere, la varietà descritta dal vettore $S(o)\,\zeta$ ha dimensione minore dell'ordine n di S e pertanto, scelti arbitrariamente n vettori linearmente indipendenti, per uno almeno di essi l'ipotesi su η è verificata.

N.B. Se λ_0 è una approssimazione di λ, posto $\lambda_0 - \lambda = t$, si ha

$\Omega - \lambda_0\,I = \Omega - \lambda\,I - t\,I$ e quindi, assumendo tale matrice come matrice $S(t)$, si può applicare il teorema precedente.

Volendo estendere il teorema precedente al caso di matrici dipendenti analiticamente da più parametri, si incontra la difficoltà consisten

A. M. Ostrowski

te nel fatto che non esiste $\quad \lim \zeta/|\zeta|$.

Ad esempio, se

$$S = S(u,v) = \begin{pmatrix} u & o & o \\ o & v & o \\ o & o & 1 \end{pmatrix} \, , \qquad \eta = (1,1,1) \, ,$$

l'equazione $S(u,v)\,\zeta = \eta$ ha, per $u\,v \neq o$, soluzione $\zeta = (1/u, 1/v, 1)$ e quindi

$$\frac{\zeta}{|\zeta|} \equiv \left(\frac{1}{\sqrt{1+u^2+u^2/v^2}} \, , \ \frac{1}{\sqrt{1+v^2+v^2/u^2}} \, , \ \frac{1}{\sqrt{1+1/u^2+1/v^2}} \right) .$$

Se u e v tendono a zero in modo che $u/v \to o$, si ha $\lim \zeta/|\zeta| = (1,0,0)$; se invece u e v tendono a zero in modo che $v/u \to 0$, si ha $\lim \zeta/|\zeta| = (0, 1, 0)$.

Esaminiamo ora il caso in cui la matrice $S = S(0)$ non sia nota, ma si conosca una matrice Ω approssimante S .

Definendo allora la matrice A in modo che $S + A = \Omega$, si pone il seguente problema: Fissato il vettore η e detta ζ_A una soluzione dell'equazione $(S + A)\,\zeta_A = \eta$, studiare il comportamento di $\zeta_A/|\zeta_A|$ quando A tende a zero, cioè quando tende a zero la norma $|A|_2$ di FRÖBENIUS della matrice A.

(Si pone $|A|_2 = \sqrt{\sum_{\mu,\nu} |a_{\mu\nu}|^2}$).

Osserviamo innanzitutto due proprietà ben note dalla norma ora introdotta, estensioni di proprietà corrispondenti nel caso dei vettori :

I) $\qquad |AB|_2 \leq |A|_2 \cdot |B|_2 \, ,$

A. M. Ostrowski

II) $\qquad |A + B|_2 \leq |A|_2 + |B|_2$.

In particolare, per un vettore ζ , si ha: $\quad |A\zeta| \leq |A|_2 |\zeta|$

Sussiste il seguente teorema :

II. <u>Se</u> S <u>è una matrice singolare,</u> A <u>una matrice tendente a zero tale che</u> S + A <u>sia non singolare, posto</u> $\zeta_A = (S + A)^{-1}\eta$, <u>esiste un vettore</u> π_A <u>tale che</u> :

(6) $\qquad S\,\pi_A = 0, \qquad\qquad \zeta_A \big/ |\zeta_A| - \pi_A = O\,(\,|A|_2) .$

<u>Dimostrazione.</u>

Dimostriamo dapprima il teorema nel caso particolare che S abbia la forma :

(7) $\qquad\qquad S \; = \begin{pmatrix} O_k & & O \\ & & \\ O & & I_{n-k} \end{pmatrix}$

ove O_k indica una matrice quadrata di ordine k nulla, I_{n-k} indica la matrice unitaria di ordine n-k e le O sono matrici rettangolari nulle.

Sia η un vettore non appartenente alla varietà descritta da S ζ ; nel caso attuale, bisogna supporre che una almeno delle prime k componenti di η sia diversa da o.

Poniamo $\eta = (y_1 , \ldots, y_n)$; $\zeta = \zeta_1 + \zeta_2$, con

$\zeta_1 = (x_1 , \ldots, x_k , \; 0, \ldots, 0)$ e $\zeta_2 = (0, \ldots, 0, x_{k+1} , \ldots, x_n)$.

Si ha :

$\qquad (S + A)\,\zeta = S\,\zeta + A\,\zeta = A\,\zeta + \zeta_2 .$

A. M. Ostrowski

e pertanto il sistema :

$$(8) \qquad (S+A)\,\zeta \;=\; \eta$$

può mettersi nella forma

$$(9) \qquad A\zeta + \zeta_2 = \eta$$

Indichiamo con χ un indice fra 1 e k. Si ha:

$$y_\chi = \sum_{\nu=1}^{n} a_{\chi\nu}\, x_\nu \,,$$

da cui, per la diseguaglianza di Cauchy-Schwartz, si ricava :

$$\left| y_\chi \right| \le \left| A \right|_2 \left| \zeta \right|$$

e, ricordando che, per almeno un valore di χ , $y_\chi \ne 0$, si deduce

$$\frac{1}{\left| \zeta \right|} \le \frac{\left| A \right|_2}{\left| y_\chi \right|} \;;$$

quindi

$$^1/_{\left| \zeta \right|} = O(\left| A \right|_2), \quad \lim \left| \zeta \right| = +\infty \,.$$

Dividendo la relazione (9) per $\left| \zeta \right|$, si ottiene :

$$\frac{A\zeta}{\left| \zeta \right|} + \frac{\zeta_2}{\left| \zeta \right|} = \frac{\eta}{\left| \zeta \right|}$$

e pertanto, essendo $\left| A\zeta \right| \le \left| A \right|_2 \left| \zeta \right|$, riesce :

A. M. Ostrowski

$$\frac{|\zeta_2|}{|\zeta|} \leq \frac{|\eta|}{|\zeta|} + \frac{1}{|\zeta|}|A|_2|\zeta| \leq \frac{|\eta|}{|\zeta|} + |A|_2 = O(|A|_2).$$

Ne segue $\quad \lim \dfrac{|\zeta_2|}{|\zeta|} = ,0, \quad \lim|\zeta_1| = +\infty .$

D'altra parte:

$$\frac{\zeta}{|\zeta|} = \frac{\zeta_1}{|\zeta|} + \frac{\zeta_2}{|\zeta|} = \frac{\zeta_1}{|\zeta|} + O(|A|_2) =$$

$$= \frac{\zeta_1}{|\zeta_1|} \cdot \frac{|\zeta_1|}{|\zeta|} + O(|A|_2),$$

$$\frac{|\zeta_1|^2}{|\zeta|^2} = 1 - \frac{|\zeta_2|^2}{|\zeta|^2} = 1 + O(|A|_2^2),$$

e quindi:

$$\zeta/|\zeta| = \zeta_1/|\zeta_1| + O(|A|_2).$$

Posto $\Pi_A = \zeta_1/|\zeta_1|$, si ha $S \Pi_A = 0$ e di conseguenza la tesi.

Per ottenere il teorema nel caso generale, cioè per un'arbitraria matrice S di rango $n - k$, basta osservare che, ammesso il teorema per una particolare matrice S_1, il teorema seguita a sussistere per ogni matrice S del tipo $S = B S_1 C$, con B, C arbitrarie matrici non singolari.

A. M. Ostrowski

Lezione II

Metodo del quoziente di RAYLEIGH.

Sia A una matrice reale simmetrica e ξ un vettore-colonna. Introdotto il vettore ξ' trasposto di ξ , si consideri la forma quadratica

$$Q_A (\xi) = \xi' A \xi = \sum_{\mu, \nu} a_{\mu \nu} x_\mu x_\nu \text{ ove } A = (a_{\mu \nu}),$$

$$\xi' = (x_1, \ldots, x_n).$$

Chiamasi quoziente di RAYLEIGH l'espressione :

$$(10) \qquad R_A (\xi) = \frac{Q_A (\xi)}{|\xi|^2}.$$

Il quoziente di RAYLEIGH gode della proprietà che, se ξ è una autosoluzione della matrice   A   relativa all'autovalore λ , riesce :

$$R_A (\xi) = \lambda.$$

Il quoziente di RAYLEIGH permette, in generale, la costruzione di una successione $\{\lambda_\nu\}$ convergente all'autovalore λ e di una successione di vettori $\{\xi_\nu\}$ convergente ad un'autosoluzione ξ relativa a λ . Sia λ_0 un'approssimazione di λ . Fissato un η generico costruiamo il vettore ξ_0 soluzione del sistema :

$$(11) \qquad (A - \lambda_0 I) \xi_0 = \eta.$$

Il quoziente di RAYLEIGH relativo a ξ_0 fornisce un numero

A. M. Ostrowski

λ_1 che protrebbe migliorare l'approssimazione di λ . Iterando il procedimento si ottengono le successioni $\{\lambda_\nu\}$ e $\{\xi_\nu\}$ di cui sopra. Faremo ora vedere che, se λ_0 è abbastanza vicino a λ , abbiamo la convergenza di $\{\lambda_\nu\}$ e questa è una convergenza <u>quadratica</u>, nel senso che esiste finito il

$$(12) \qquad \lim_{\nu \to \infty} \frac{\lambda_\nu - \sigma}{\left|\lambda_{\nu-1} - \sigma\right|^2} \ .$$

Poichè, come è evidente, il quoziente di RAYLEIGH è invariante rispetto alle trasformazioni S ortogonali $(SS' = I)$, e poichè una matrice simmetrica può essere diagonalizzata con una trasformazione ortogonale, supponiamo senz'altro $A = \mathrm{diag}\,(\mu_1, \ldots, \mu_n)$. I numeri μ_k non sono necessariamente distinti; indichiamo allora con $\sigma_1, \ldots, \sigma_m$ quelli dei $\mu_1, \ldots, \mu_n$ che sono fra loro distinti e supponiamo che $\mu_1 = \ldots = \mu_h = \sigma_1$ e $\mu_k \neq \sigma_1$ per $k > h$.

Si ha allora :

$$(13) \qquad R_A\,(\xi) = \frac{\sum_{k=1}^{n} \mu_k |x_k|^2}{\sum_{k=1}^{n} |x_k|^2} \ .$$

Indichiamo con $\eta = (y_1, \ldots, y_n)$ un vettore avente diversa da zero almeno una delle prime h componenti; riesce:

$$\xi_0 = \left(\frac{y_1}{\mu_1 - \lambda_0} , \ldots , \frac{y_n}{\mu_n - \lambda_0} \right) .$$

Dalla (13) si deduce :

A. M. Ostrowski

$$R_A(\xi_0) = \frac{\displaystyle\sum_{k=1}^{n} \mu_k \frac{|y_k|^2}{(\mu_k - \lambda_0)^2}}{\displaystyle\sum_{k=1}^{n} \frac{|y_k|^2}{(\mu_k - \lambda_0)^2}} = \frac{\displaystyle\sum_{k=1}^{m} \sigma_k \frac{p_k}{|\sigma_k - \lambda_0|^2}}{\displaystyle\sum_{k=1}^{m} \frac{p_k}{|\sigma_k - \lambda_0|^2}},$$

ove si è posto $\quad p_1 = \displaystyle\sum_{k=1}^{h} |y_k|^2 \quad$ e analogamente per $p_2, \ldots, p_m$.

Poniamo ora $\sigma = \sigma_1$, $p = p_1$. Si ha :

$$(14) \qquad R_A(\xi_0) = \frac{\displaystyle\sum_{k=1}^{m} \sigma_k \frac{p_k}{|\sigma_k - \lambda_0|^2}}{\dfrac{p}{|\sigma - \lambda_0|^2} + \displaystyle\sum_{k=2}^{m} \frac{p_k}{|\sigma_k - \lambda_0|^2}}.$$

Posto $R_A(\xi_0) = \lambda_1$, dalla (14) si trae :

$$(15) \qquad \lambda_1 - \sigma = \frac{\displaystyle\sum_{k=2}^{m} (\sigma_k - \sigma) \frac{p_k}{|\sigma_k - \lambda_0|^2}}{\dfrac{p}{|\sigma - \lambda_0|^2} + \displaystyle\sum_{k=2}^{m} \frac{p_k}{|\sigma_k - \lambda_0|^2}} = \frac{\displaystyle\sum_{k=2}^{m} (\sigma_k - \sigma) \frac{p_k}{|\sigma_k - \lambda_0|^2}}{p + \displaystyle\sum_{k=2}^{m} \frac{p_k |\sigma - \lambda_0|^2}{|\sigma_k - \lambda_0|^2}} |\sigma - \lambda_0|^2.$$

Posto :

$$L = \frac{\displaystyle\sum_{k=2}^{m} p_k \frac{\sigma_k - \sigma}{|\sigma_k - \sigma|^2}}{p}$$

A. M. Ostrowski

dalla (15) si trae che, se λ_0 è sufficentemente prossimo a σ ,

$$\frac{\lambda_1 - \sigma}{|\lambda_0 - \sigma|^2}$$ è abbastanza prossimo a L.

Iterando il ragionamento, si avrà allora $\lim_{\nu \to \infty} \lambda_\nu = \sigma$ e

$$\lim \frac{\lambda_{\nu+1} - \sigma}{|\lambda_\nu - \sigma|^2} = L.$$

Per quanto riguarda l'approssimazione dell'autosoluzione si ha:

$$\xi_\nu = \left(y_1, \ldots, y_h, y_{h+1} \frac{\sigma - \lambda_\nu}{\mu_{h+1} - \lambda_\nu}, \ldots, y_n \frac{\sigma - \lambda_\nu}{\mu_n - \lambda_\nu} \right),$$

e, per $\lambda_\nu \to \sigma$, riesce $\lim \xi_\nu = \pi$, avendo posto $\pi = (y_1, \ldots, y_h, 0,$

$0, \ldots, 0)$; π è un'autosoluzione relativa a σ . Riesce

$$(16) \quad \xi_\nu - \pi = (\sigma - \lambda_\nu)(0, \ldots, 0, \frac{y_{h+1}}{\mu_{h+1} - \lambda_\nu}, \ldots, \frac{y_n}{\mu_n - \lambda_\nu}) =$$

$$= (\sigma - \lambda_\nu) \pi_1 + O(|\sigma - \lambda_\nu|^2),$$

ove si è posto $\pi_1 = \left(0, \ldots, 0, \frac{y_{h+1}}{\mu_{h+1} - \sigma}, \ldots, \frac{y_n}{\mu_n - \sigma} \right).$

Quanto sopra detto si estende in modo quasi evidente al caso delle matrici hermitiane. Naturalmente in questo caso $Q_A(\xi)$ è definito come $\xi^* A \xi$ e la diagonalizzazione si ottiene usando invece delle matrici ortogonali, le matrici unitarie ortogonali S definite dalla relazione $S^* S = I$ (S^* è generalmente la matrice coniugata dalla trasposta di S).

A. M. Ostrowski

Più in generale, la nostra argomentazione si applica anche al caso delle **matrici normali**, cioè di quelle matrici che possono essere diagonalizzate per mezzo di una trasformazione unitaria ortogonale; gli autovalori di una matrice normale non sono però necessariamente reali. Le matrici normali possono anche essere definite per mezzo della relazione $AA^* = A^* A$, che mostra che l'insieme di queste matrici è un'intersezione delle varietà quadratiche nella geometria hermitiana (nel campo di tutte le matrici complesse). Un'altra caratterizzazione delle matrici normali A, dovuta a I. SCHUR, consiste nella proprietà che $|A|_2 = \sqrt{|\lambda_1|^2 + |\lambda_2|^2 + \dots |\lambda_m|^2}$, ove $\lambda_1, \dots, \lambda_m$ sono gli autovalori di A. La teoria delle matrici normali, dovuta quasi nella sua totalità a I. SCHUR, si presenta come un'estensione elegantissima della teoria delle matrici hermitiane; lavorando con le matrici normali, bisogna avere particolare cura, perchè l'insieme di dette matrici non è lineare, fatto che facilmente si dimentica.

Metodo di convergenza cubica.

Modificando il procedimento ricorrente al quale abbiamo accennato e sostituendo ad ogni passo ad η il vettore già formato $\xi_{\nu-1}$ è possibile costruire una successione di numeri $\{\lambda_\nu\}$ e una successione di vettori unitari $\{\xi_\nu\}$ tali che :

$$(17) \qquad \lambda_{\nu+1} = R_A(\xi_\nu) \quad , \qquad \xi_\nu = (A - \lambda_\nu I)\,\xi_{\nu-1} \ .$$

A. M. Ostrowski

Questo procedimento presenta una convergenza cubica. La prima discussione di questo fatto, dovuta a CRANDALL [3], è la seguente : Supposta la matrice A diagonalizzata e fissato l'indice ν , si moltiplichi il vettore ξ_ν per un fattore, in modo che le prime h componenti non dipendano da ν . Indicando tale vettore con ζ_ν , scriveremo :

$$(18) \qquad \zeta_\nu = \quad (y_1, \ldots, y_n, \ z_{h+1}^{(\nu)} \ , \ldots, \ z_n^{(\nu)}).$$

Introduciamo il vettore $\Pi = (y_1, \ldots, y_h, 0, \ldots, 0)$ e indichiamo con ε il massimo fra $\left| z_{h+1}^{(\nu)} \right| , \ldots, \left| z_n^{(\nu)} \right|$.

Riesce ovviamente

$$\zeta_\nu - \Pi = \quad O(\varepsilon).$$

Passando all'indice $\nu + 1$ e indicando con $x_1, \ldots, x_n$ le componenti di $\xi_{\nu+1}$ si ottiene :

$$x_i = \frac{y_i}{\sigma - \lambda_{\nu+1}} \text{ per } i = 1, \ldots, h, \ x_k = \frac{z_k^{(\nu)}}{\mu_k - \lambda_{\nu+1}} \text{ per } k > h,$$

da cui :

$$\xi_{\nu+1} = \frac{\Pi}{\sigma - \lambda_{\nu+1}} + \left(0, \ldots, 0, \ \frac{z_{h+1}^{(\nu)}}{\mu_{h+1} - \lambda_{\nu+1}}, \ldots, \frac{z_n^{(\nu)}}{\mu_n - \lambda_{\nu+1}} \right);$$

---

(3) Cfr. S.H. CRANDALL, Iterative procedures related to relaxation methods for eigenvalue problems, Proc. Roy. Soc. London, 207, 1951, 421 - 422.

A. M. Ostrowski

moltiplicando il vettore $\xi_{\nu+1}$ per il fattore $\sigma - \lambda_{\nu+1}$ si ricava:

$$(19) \qquad \zeta_{\nu+1} = \Pi + \left(0,\dots,0,\ \frac{z_{h+1}^{(\nu)}}{\mu_{h+1}-\lambda_{\nu+1}},\dots,\frac{z_n^{(\nu)}}{\mu_m-\lambda_{\nu+1}}\right)(\sigma - \lambda_{\nu+1}).$$

Ricordiamo ora che riesce

$$R(\zeta_\nu) - \sigma = \lambda_{\nu+1} - \sigma = \frac{\displaystyle\sum_{k=h+1}^{m}(\mu_k - \sigma)\left|z_k^{(\nu)}\right|^2}{\left|\zeta_\nu\right|^2}$$

e quindi, essendo $\left|z_k^{(\nu)}\right| \le \varepsilon$, si ottiene :

$$\lambda_{\nu+1} - \sigma = O(\varepsilon^2)$$

Introducendo questa espressione in (19) si ha :

$$(20) \qquad \zeta_{\nu+1} - \Pi = O(\varepsilon^3).$$

Questa argomentazione è piuttosto un'argomentazione di plausibilità e non mette in luce che anche la convergenza di $\left\{\lambda_\nu\right\}$ è cubica, cioè che riesce :

$$(21) \qquad \lambda_{\nu+1} - \sigma = O\left[(\lambda_\nu - \sigma)^3\right].$$

Usando la terminologia dell'amico WEINSTEIN, bisogna dunque dire che in questo momento il matematico di alt o ingegno deve cedere il pas - so al "farmacista". La discussione rigorosa del metodo permette effettivamente di provare non soltanto la (21), ma anche che

$$(22) \qquad \lim_{\nu \to \infty} \frac{\lambda_{\nu+1} - \sigma}{(\lambda_\nu - \sigma)^3} = g > 0.$$

A. M. Ostrowski

La dimostrazione di questo risultato, con le indicazioni di tutte le
ipotesi da fare, sarà l'argomento delle lezioni IV e V.

A. M. Ostrowski

Lezione III

Quoziente di RAYLEIGH generalizzato. (Argomentazioni di plausibilità).

Sia A una matrice non necessariamente hermitiana; ad essa si può associare una forma bilineare $\eta\, A\, \xi$.

Si definisce allora come quoziente di RAYLEIGH l'espressione:

$$(23) \qquad R_A(\xi,\eta) = \frac{\eta\, A\, \xi}{\eta\, \xi}, \qquad\qquad \eta\, \xi \neq 0$$

(d'ora in poi i vettori si intenderanno come <u>vettori riga</u> se sono fattori di sinistra e <u>vettori colonna</u> se sono fattori di destra). Si possono stabilire le formule ricorrenti :

$$(24) \qquad \lambda_{\nu+1} = R_A(\xi_\nu,\eta_\nu)$$

$$(25) \qquad \xi_\nu = (A - \lambda_\nu I)^{-1}\, \xi_{\nu-1}\,, \qquad \eta_\nu = \eta_{\nu-1}(A - \lambda_\nu I)^{-1}.$$

La proprietà estremale delle autosoluzioni per il quoziente di RAYLEIGH, sulla quale si fondava il principio di massimo nel metodo di calcolo dell'autovalore (cfr. lez. II), non sussiste più; in questo caso la proprietà suddetta può essere sostituita da una proprietà di <u>stazionarie-tà</u>[4].

---

(4) Cfr. A. M. OSTROWSKI, <u>On the convergence of the Rayleigh quotient Iteration for the computation of the Characteristic Roots on Vectors</u> III. Archive Rat. Mech. and Anal. vol. 3, 1959, 326.

A. M. Ostrowski

Ciò, ad esempio, si verifica quando all'autovalore corrispondono soltanto i divisori elementari lineari della matrice [5].

In effetti, se σ è un autovalore di molteplicità k cui
corrispondono soltanto divisori elementari lineari, esiste una matrice
S non degenere tale che

$$S^{-1} A S = \sigma I_k + D_{n-k} = \begin{pmatrix} \sigma I_k & 0 \\ 0 & D_{n-k} \end{pmatrix}$$

ove I_k è la matrice identica di ordine k e D_{n-k} è una matrice quadrata di
ordine n-k i cui autovalori sono tutti distinti da σ .

Sia ξ (η) un'autosoluzione a destra (a sinistra) corrispon-
dente a σ : $\xi = (a_1, \ldots a_k, 0, \ldots, 0)$, $\eta = (b_1, \ldots, b_k, 0, \ldots, 0)$;

poniamo $p = \eta \xi = \sum_{\nu=1}^{k} a_\nu b_\nu$ <u>e supponiamolo diverso da 0.</u>

Siano ora ξ_1 e η_1 due vettori prossimi rispettivamente a ξ e η, supposti nella forma :

$$\xi_1 = (a_1, \ldots, a_k, \alpha_{k+1}, \ldots, \alpha_n), \quad \eta_1 = (b_1, \ldots, b_k, \beta_{k+1}, \ldots, \beta_n),$$

essendo le α_i, $\beta_i = O(\varepsilon)$ per i = k+1,...,n. Si ha :

---

(5) Un'altra via conducente al metodo descritto sopra è stata data da
M. R. HESTENES, <u>Inversion of matrices by biortogonalization and
related results</u>, J. Soc. Indust. Appl. Math., vol. 6, 1958,
80 - 83.

A. M. Ostrowski

$$\eta_1 \xi_1 = p + \sum_{i=k+1}^{n} \alpha_i \beta_i = p + O(\varepsilon^2),$$

$$\eta_1 A \xi_1 = \sigma p + O(\varepsilon^2).$$

Pertanto :

$$R_A(\xi_1, \eta_1) = \frac{\sigma p + O(\varepsilon^2)}{p + O(\varepsilon^2)} = \sigma + O(\varepsilon^2).$$

In quest'ultima espressione si riconosce una certa "stazionarietà" del quoziente generalizzato[6].

Mostriamo ora, con un esempio, che, se non è verificata la ipotesi precedente, non si ha in generale stazionarietà. Sia:

$$A = \begin{pmatrix} \sigma & 1 \\ 0 & \sigma \end{pmatrix}, \quad \xi = (x_1, x_2)', \quad \eta = (y_1, y_2).$$

Si ottiene allora :

$$R_A(\xi, \eta) = \frac{\sigma(x_1 y_1 + x_2 y_2) + x_2 y_1}{x_1 y_1 + x_2 y_2} = \sigma + \frac{x_2 y_1}{x_1 y_1 + x_2 y_2}.$$

In questo caso σ è autovalore e un'autosoluzione a sinistra è (0, 1) e

____

(6) Cfr. Nota citata in [4] pag. 327.

A. M. Ostrowski

e a destra è $(1, 0)'$.

Posto $\xi' = (1, \alpha)$ e $\eta = (\beta, 1)$, riesce :

$$R_A(\xi, \eta) = \sigma + \frac{\alpha \beta}{\alpha + \beta}.$$

Da ciò si trae l'asserto.

Altre generalizzazioni del quoziente di RAYLEIGH, in una nuova direzione, si ottengono dall'osservare che $A - \lambda I$ è una particolare matrice dipendente linearmente dal parametro λ ; sostituendo allora $A - \lambda I$ con la più generale matrice dipendente linearmente da λ , si ottiene il problema di autovalori :

$$(A_o \lambda + A_1) \xi = 0.$$

Per questo problema si può sviluppare una teoria analoga a quella precedente, definendo il quoziente di RAYLEIGH al modo seguente:

$$R_A(\xi, \eta) = - \frac{\eta A_1 \xi}{\eta A_o \xi} \tag{7}$$

Un'ulteriore generalizzazione si ottiene sostituendo ad una matrice lineare in λ una matrice del tipo :

$$D(\lambda) = A_o \lambda^m + A_1 \lambda^{m-1} + \ldots + A_{m-1} \lambda + A_m, \text{ ove le } A_i$$

$(i = 0, \ldots, m)$ sono matrici quadrate di ordine n.

---

(7) Cfr. S.H. CRANDALL, Iterative procedure ....., Proc. Roy. Soc. London, 1951, 417.

A. M. Ostrowski

In questo caso un modo di definire il quoziente di RAYLEIGH, dovuto a P. LANCASTER[8], è il seguente :

$$R_D (\xi , \eta , \lambda) = \lambda - \frac{\eta \, D(\lambda) \, \xi}{\eta \, D'(\lambda) \, \xi},$$

ove si è posto $D'(\lambda) = m \, A_0 \, \lambda^{m-1} + (m-1) \, A_1 \, \lambda^{m-2} + \ldots + A_{m-1}.$

Per mezzo di tale quoziente si può definire un procedimento iterativo, mediante le formule :

$$\lambda_{\nu+1} = \lambda_\nu - \frac{\eta_\nu \, D(\lambda_\nu) \, \xi_\nu}{\eta_\nu \, D'(\lambda_\nu) \, \xi_\nu},$$

$$\xi_{\nu+1} = \left[D(\lambda_{\nu+1}) \right]^{-1} \xi_\nu \, , \qquad \eta_{\nu+1} = \eta_\nu \left[D(\lambda_{\nu+1}) \right]^{-1}.$$

Osserviamo i seguenti casi particolari:

per $m=1$, si ha: $\quad R_D(\xi , \eta , \lambda) = - \dfrac{\eta \, A_1 \, \xi}{\eta \, A_0 \, \xi} \, ,$

che rappresenta la prima generalizzazione introdotta in questa lezione;

per $n = 1$, A_i sono scalari e $D(\lambda)$ è un polinomio; il procedimento ricorrente introdotto si rappresenta con la formula:

---

(8) Cfr. P. LANCASTER, A Generalized Rayleigh Quotient Iteration for Lambda - Matrices, Archive Rat. Mech. and Anal. VIII, 1961, 309-310.

A. M. Ostrowski

$$\lambda_{\nu+1} = \lambda_\nu - \frac{D(\lambda_\nu)}{D'(\lambda_\nu)}$$

e coincide pertanto con il ben noto procedimento di NEWTON.

Pare dapprima che nel caso delle matrici quadrate A qualsiasi, il quoziente di RAYLEIGH generalizzato è il solo razionalmente applicabile. In qualche calcolo eseguito all'Istituto di Calcolo di Ramo - Wooldridge Corporation in Los Angeles, il metodo del quoziente classico di RAYLEIGH era applicato anche alle matrici complesse non hermitiane, con un successo sorprendente. Lo studio dettagliato di questo caso mostra che c'è generalmente una convergenza, ma soltanto quadrática. Questo è però compensato dal fatto che, utilizzando il quoziente classico, bisogna ad ogni passo risolvere soltanto un sistema lineare anzichè due, come nel caso del quoziente generalizzato.

Inoltre è molto facile lavorare con il denominatore $\xi^* \xi$ invece del prodotto $\eta \, \xi$ che può tendere a zero[9] molto rapidamente.

---

(9) Cfr. A. M. OSTROWSKI, On the Convergence of the Rayleigh Quotient ....., V, Archive Rat. Mech. and Anal. III, 1959, ed ancora A. M. OSTROWSKI, On the Convergence of the Rayleigh Quotient ...., VI, Archive Rat. Mech. and Anal., IV, 1959.

A. M. Ostrowski

Lezione IV

La legge asintotica per il metodo del Quoziente di RAYLEIGH

Dimostriamo ora il seguente teorema :

III. Se S è una matrice hermitiana, σ un suo autovalore, $\{\lambda_\nu\}$ una successione, costruita per mezzo del quoziente di RAYLEIGH, approssimante σ , e tale che $\lambda_\nu \neq \sigma$ per ogni ν , sussiste la seguente proprietà:

$$(26) \qquad \lim_{\nu \to \infty} \frac{\lambda_{\nu+1} - \sigma}{\left| \lambda_\nu - \sigma \right|^2 \left(\lambda_\nu - \sigma \right)} = \gamma > 0.$$

Se S è normale e non hermitiana, la (26) sussiste sotto l'ulteriore condizione che non esistano due autovalori σ_i e σ_i' distinti fra loro e da σ , aventi da σ la stessa distanza.

Dimostrazione. [10]

Sia $A = \operatorname{diag}(\mu_1 , \ldots , \mu_m)$ e $\sigma, \sigma_1 , \ldots , \sigma_m$ i valori distinti di $\mu_1, \mu_2 , \ldots , \mu_n$; supponiamo $\sigma = \mu_1 = \ldots = \mu_h$ e $\mu_h \neq \mu_k$ se $k > h$. Consideriamo un vettore $\eta = (y_1 , \ldots , y_n)$ avente almeno una delle prime h componenti diversa da zero e poniamo $\xi_0 = \eta$. Mediante il procedimento ricorrente esposto nelle lezioni precedenti, si può formare una successione di numeri $\{\lambda_\nu\}$ che sono supposti approssimanti l'autovalore σ e costruire in corrispondenza ad essi la successio-

---

(10) Cfr. A. M. OSTROWSKI, On the convergence of the Rayleigh Quotient .... I, ARCHIVE Rat. Mech. and Anal., vol. I, 1958, 233-241.

A. M. Ostrowski

ne di vettori $\left\{\xi_\nu\right\}$ definiti dalle relazioni :

$$(27) \qquad x_k^{(\nu+1)} = \frac{x_k^{(\nu)}}{\mu_k - \lambda_{\nu+1}} \qquad (k = 1, \ldots, n),$$

avendo posto $\xi_\nu = (x_1^{(\nu)}, \ldots, x_n^{(\nu)})$.

Dalle (27) si ricava, ricordando che $\xi_0 = \eta$:

$$(28) \qquad x_k^{(\nu)} = \frac{y_k}{\prod_{\tau=1}^{\nu}(\mu_k - \lambda_\tau)}.$$

Poniamo :

$$N_\nu = \prod_{\tau=1}^{\nu}(\sigma - \lambda_\tau), \qquad N_{\nu,k} = \prod_{\tau=1}^{\nu}(\sigma_k - \lambda_\tau), \qquad p = y_1^2 + \ldots + y_h^2 > 0;$$

riesce allora :

$$|\xi_\nu|^2 = \frac{p}{|N_\nu|^2} + \sum_{k=1}^{m} \frac{p_k}{|N_{\nu,k}|^2},$$

$$\xi_\nu^* A\,\xi_\nu = \frac{p\,\sigma}{|N_\nu|^2} + \sum_{k=1}^{m} \frac{p_k\,\sigma_k}{|N_{\nu,k}|^2})$$

denotando p_k opportuni numeri positivi. Ne segue che :

A. M. Ostrowski

$$\lambda_{\nu+1} - \sigma \;=\; \frac{\xi_\nu^* \, A \, \xi_\nu}{|\xi_\nu|^2} - \sigma \;=\; \frac{\displaystyle\sum_{k=1}^{m} \frac{p_k\,(\sigma_k - \sigma)}{|N_{\nu,k}|^2}}{\dfrac{p}{|N_\nu|^2} + \displaystyle\sum_{k=1}^{m} \frac{p_k}{|N_{\nu,k}|^2}} \, ,$$

e quindi

$$\frac{\lambda_{\nu+1} - \sigma}{|N_\nu|^2} \;=\; \frac{Z_\nu}{K_\nu} \, ,$$

ove si ponga

$$Z_\nu \;=\; \sum_{k=1}^{m} p_k \,(\sigma_k - \sigma)\, \frac{1}{|N_{\nu,k}|^2},$$

$$K_\nu \;=\; p + \sum_{k=1}^{m} p_k \left| \frac{N_\nu}{N_{\nu,k}} \right|^2.$$

Facciamo ora vedere che, se si prende λ_1 abbastanza vicino a σ , allora la successione $\{\lambda_\nu\}$, converge a σ . Sia $c_1 = \frac{1}{2}\,\min\left\{\,|\sigma_1 - \sigma|\,,\ldots,\,|\sigma_m - \sigma|\,\right\}$ e c un numero positivo minore di c_1. Supponiamo che $|\sigma - \lambda_\tau| \leqq c < c_1,$ per $\tau = 1,\ldots,\nu$. Riesce allora :

$$|N_{\nu,k}| \;=\; \prod_{\tau=1}^{\nu} |\sigma_k - \lambda_\tau| > c_1^\nu \, ,$$

A. M. Ostrowski

$$\left| z_{\nu} \right| \leq \frac{1}{c_1^{2\nu}} \sum_{k=1}^{m} p_k \left| \sigma_k - \sigma \right| = \frac{c_2}{c_1^{2\nu}},$$

avendo posto
$$c_2 = \sum_{k=1}^{m} p_k \left| \sigma_k - \sigma \right|.$$

Dalle disuguaglianze precedenti e dall'osservazione che $K_{\nu} \geqslant p$, segue :

$$\frac{\left| \lambda_{\nu+1} - \sigma \right|}{\left| N_{\nu} \right|^2} \leq \frac{c_2}{p\, c_1^{2\nu}}$$

e quindi

$$(29) \qquad \frac{\left| \lambda_{\nu+1} - \sigma \right|}{c} \leq \frac{c_2}{p\, c} \quad \frac{\left| N_{\nu} \right|^2}{c_1^{2\nu}} \leq \frac{c_2}{p\, c} \left(\frac{c}{c_1} \right)^{2\nu}.$$

Scegliamo c in modo che sia positivo e minore dei numeri c_1, $\dfrac{p\, c_1^2}{c_2}$; dalla (29), per $\nu = 1, 2, \ldots$ si ricava $\left| \lambda_{\nu+1} - \sigma \right| \leq c$ ed ancora, posto $c_3 = c_2/p$:

A. M. Ostrowski

$$\left| \lambda_{\nu+1} - \sigma \right| \leq c_3 \left(\frac{c}{c_4} \right)^{2\nu} \ ; $$

questa relazione è dunque vera per ogni ν se si verifica per $\nu = 1$;

ciò prova la convergenza della successione $\left\{ \lambda_\nu \right\}$ a σ

Osserviamo ora il seguente :

Lemma I.   Se $\left| \lambda_1 - \sigma \right| < c$,   allora

$$(30) \qquad \lambda_{\nu+1} - \sigma = O\left((\lambda_\nu - \sigma)^2 \right)$$

Si ha infatti dalla (29) :

$$\frac{\left| \lambda_{\nu+1} - \sigma \right|}{\left| \lambda_\nu - \sigma \right|} - \frac{c_3}{c_4^{2\nu}} \prod_{\tau=1}^{\nu-1} \left| \lambda_\nu - \sigma \right|^2 = \frac{c_3}{c_4^2} \prod_{\tau=1}^{\nu-1} \left| \frac{\lambda_\tau - \sigma}{c_4} \right|^2 < \frac{c_3}{c_4^2} \ .$$

Dal Lemma I seguono immediatamente i seguenti:

Corollario 1. Se c_4 è una costante arbitraria, allora

$$(31) \qquad \lim_{\nu \to \infty} c_4^\nu (\lambda_\nu - \sigma) = 0.$$

Si ha infatti :

A. M. Ostrowski

$$\frac{c_4^{\nu+1}(\lambda_{\nu+1}-\sigma)}{c_4^{\nu}(\lambda_{\nu}-\sigma)} = O(\lambda_{\nu}-\sigma).$$

<u>Corollario 2.</u>   <u>La serie $\displaystyle\sum_{\nu=1}^{\infty}|\lambda_{\nu}-\sigma|$ è convergente.</u>

Si ha infatti :

$$\left|\frac{\lambda_{\nu+1}-\sigma}{\lambda_{\nu}-\sigma}\right| = O(\lambda_{\nu}-\sigma).$$

<u>Lemma II.</u>   <u>Per ogni indice k, esiste un numero T_k tale</u> che:

$$(32)\qquad \frac{1}{N_{\nu,k}} = \frac{1}{(\sigma_k-\sigma)^{\nu}}\left[T_k + O(\lambda_{\nu+1}-\sigma)\right].$$

Dimostrazione.

Per quanto procede, si ha :

$$N_{\nu,k} = \prod_{\tau=1}^{\nu}(\sigma_k-\sigma+\sigma-\lambda_{\tau}) = (\sigma_k-\sigma)^{\nu}\prod_{\tau=1}^{\nu}\left(1+\frac{\sigma-\lambda_{\tau}}{\sigma_k-\sigma}\right).$$

Poichè la serie $\displaystyle\sum_{\tau=1}^{\infty}\frac{\sigma-\lambda_{\tau}}{\sigma_k-\sigma}$ è assolutamente convergente (cfr. corollario 2), posto

A. M. Ostrowski

$$T_{\nu,k} = \prod_{\tau=1}^{\nu} \left(1 + \frac{\sigma - \lambda_\tau}{\sigma_k - \sigma} \right) ,$$

esiste finito e diverso da zero il limite delle successioni $T_{\nu,k}$ per $\nu \to \infty$. Posto $t_k = \lim_{\nu \to \infty} T_{\nu,k}$, si ha :

$$(33) \qquad t_k - T_{\nu,k} = \sum_{\mu=\nu}^{\infty} \left(T_{\mu+1,k} - T_{\mu,k} \right) .$$

Riesce inoltre

$$T_{\mu+1,k} - T_{\mu,k} = T_{\mu,k} \left(1 + \frac{\sigma - \lambda_{\mu+1}}{\sigma_k - \sigma} - 1 \right)$$

donde

$$(34) \qquad \left| T_{\mu+1,k} - T_{\mu,k} \right| \leq c_5 \left| \sigma - \lambda_{\mu+1} \right|$$

ove c_5 è un numero maggiore di tutti i numeri $\dfrac{T_{\mu,k}}{|\sigma_k - \sigma|}$.

Dalle (33) e (34) segue :

$$\left| t_k - T_{\nu,k} \right| \leq c_5 \left| \sigma - \lambda_{\nu+1} \right| \left(1 + \sum_{\mu=\nu+1}^{\infty} \left| \frac{\sigma - \lambda_{\mu+1}}{\sigma - \lambda_{\nu+1}} \right| \right) .$$

Per il Lemma I esiste una costante c_6 tale che :

A. M. Ostrowski

$$\left| \sigma - \lambda_{\mu+1} \right| \leq c_6 \left| \sigma - \lambda_{\mu} \right|^{2}$$

e quindi, per ν abbastanza grande :

$$(35) \qquad \frac{\left| \sigma - \lambda_{\mu+1} \right|}{\left| \sigma - \lambda_{\nu+1} \right|} \leq c_6 \frac{\left| \sigma - \lambda_{\mu} \right|}{\left| \sigma - \lambda_{\nu+1} \right|} \leq c_6 \left| \sigma - \lambda_{\mu} \right|.$$

Dalla (35) segue la convergenza della serie $\displaystyle\sum_{\mu=\nu+1}^{\infty} \left| \frac{\sigma - \lambda_{\mu+1}}{\sigma - \lambda_{\nu+1}} \right|$ e pertanto esiste una costante c_7 tale che :

$$\left| t_k - T_{\nu,k} \right| \leq c_7 \left| \sigma - \lambda_{\nu+1} \right|.$$

Si ha infine :

$$N_{\nu,k} = \left(\sigma_k - \sigma \right)^{\nu} \left[t_k + O\left(\sigma - \lambda_{\nu+1} \right) \right]$$

e quindi

$$\frac{\left(\sigma_k - \sigma \right)^{\nu}}{N_{\nu,k}} = \frac{1}{t_k + O\left(\sigma - \lambda_{\nu+1} \right)} = \frac{1}{t_k} + O\left(\sigma - \lambda_{\nu+1} \right).$$

E' così provato il Lemma II.

A. M. Ostrowski

Lezione V

Fine della dimostrazione della legge asintotica. Vari metodi per accelerare la convergenza.

Applichiamo ora i Lemmi I e II per ottenere una rappresentazione asintotica di Z_ν

Si ha :

$$Z_\nu = \sum_{k=1}^{m} p_k \frac{\sigma_k - \sigma}{\left|N_{\nu,k}\right|^2} = \sum_{k=1}^{m} \frac{\alpha_k}{\left|\sigma_k - \sigma\right|^{2\nu}} + O\left(\lambda_{\nu+1} - \sigma\right),$$

ove α_k sono opportune costanti.

Indichiamo con $\gamma_1 , \ldots , \gamma_s$ i valori distinti di $\dfrac{1}{\left|\sigma_k - \sigma\right|^2}$

$(k = 1, \ldots, m)$;   si ha allora per Z_ν l'espressione :

$$(36) \quad Z_\nu = \sum_{i=1}^{s} b_i \gamma_i^{\nu} + O\left(\lambda_{\nu+1} + \sigma\right).$$

Consideriamo ora l'espressione

$$K_\nu = p + \sum_{k=1}^{m} p_k \frac{\left|N_\nu\right|^2}{\left|N_{\nu,k}\right|^2} ;$$

A. M. Ostrowski

poichè riesce

$$\frac{N_\nu}{N_{\nu,k}} = \prod_{\tau=1}^{\nu} \left| \frac{\sigma - \lambda_\tau}{\sigma_k - \lambda_\tau} \right|^q \leq \frac{1}{c_1^{2\nu}} \prod_{\tau=1}^{\nu} (\sigma - \lambda_\tau)$$

si ha :

$$\lim_{\nu \to \infty} \frac{N_\nu}{N_{\nu,k}} = 0$$

e quindi :

$$\lim_{\nu \to \infty} K_\nu = p.$$

Possiamo allora scrivere

$$\frac{Z_\nu}{K_\nu} = \frac{\lambda_{\nu+1} - \sigma}{|N_\nu|^2} = \frac{Z_\nu}{p + o\,(1)} \ .$$

Supponiamo prima che la matrice S sia hermitiana. Si ha:

$$(37) \qquad \lambda_{\nu+1} - \sigma = \left(\frac{1}{p} + o\,(1) \right) \prod_{\tau=1}^{\nu} |\sigma - \lambda_\tau|^2 \left\{ c_1 \gamma_1^\nu + \ldots + \right.$$

$$\left. + c_\ell \gamma_\ell^\nu + O\,(\lambda_{\nu+1} - \sigma) \right\} \ ,$$

ove con $c_1, \ldots, c_\ell$ si intendono quei numeri $b_1, \ldots, b_s$ non nulli, ordinanti in modo che riesca $\gamma_1 > \gamma_2 > \ldots > \gamma_\ell$.

Supponiamo che $\ell > 0$: sia allora $c_1 \neq 0$. Possiamo scrivere :

A. M. Ostrowski

$$\lambda_{\nu+1} - \sigma \ \sim \ \frac{c_1}{p} \, \gamma_1^{\nu} \prod_{\tau=1}^{\nu} \left| \sigma - \lambda_\tau \right|^2 \quad ,$$

$$\lambda_{\nu} - \sigma \ \sim \ \frac{c_1}{p} \, \gamma_1^{\nu-1} \prod_{\tau=1}^{\nu-1} \left| \sigma - \lambda_\tau \right|^2$$

e quindi

$$\frac{\lambda_{\nu+1} - \sigma}{\lambda_\nu - \sigma} \ \sim \ \gamma_1 \left| \sigma - \lambda_\nu \right|^2 .$$

Ne **segue che**

$$\lim_{\nu \to \infty} \frac{\lambda_{\nu+1} - \sigma}{\left| \lambda_\nu - \sigma \right|^2 \left(\lambda_\nu - \sigma \right)} = \gamma_1 > 0 \ .$$

Rimane ora da dimostrare che $\ell > 0$. **Nel caso opposto dalle (37) seguireb-
bero le relazioni** :

$$\lambda_{\nu+1} - \sigma = 0 \ \left\{ \prod_{\tau=1}^{\sigma} \left| \sigma - \lambda_\tau \right|^2 \left(\lambda_{\nu+1} - \sigma \right) \right\} ,$$

dalle quali si dedurrebbe :

$$\frac{1}{\prod_{\tau=1}^{\nu} \left| \sigma - \lambda_\tau \right|^2} = O\,(1)$$

il che è impossibile.

Sia ora S **una matrice normale non hermitiana. Nella ipotesi supple-**

A. M. Ostrowski

mentare fatta in questo caso, i numeri $\dfrac{1}{|\sigma_k - \sigma|^2}$ sono tutti distinti

fra loro e pertanto si ha $\ell = s = m$; inoltre i numeri $\gamma_1 , \ldots , \gamma_m$ posso-

no essere ordinati in modo che $\gamma_1 > \gamma_2 > \ldots \ldots > \gamma_m$. Ripetendo allora

l'argomentazione del caso precedente si giunge alla tesi.

Si osservi che per mezzo di una matrice unitaria ortogonale è sempre possibile diagonalizzare la matrice S : pertanto il teorema è completamente dimostrato.

Esempio[11]

Applichiamo il metodo ora esposto alla matrice $S = \begin{pmatrix} 3 & 2 \\ 2 & 6 \end{pmatrix}$, la quale ha come autovalori 2 e 7 . Scegliendo come vettore ξ_0 il vettore $(1, 0)$, si trovano come approssimazioni del primo autovalore e delle corrispondenti autosoluzioni i numeri riportati nella seguente tabella :

i	λ_i	$x_1^{(i)}$	$x_2^{(i)}$
0	3	1	0
1	2.076923	1.5	-1
2	$2.0\overset{4}{}19073413568826$	$2.0\overset{2}{}9803921568627451$	-1
4	$2.0\overset{15}{}277$	$1.9\overset{7}{}62747097570$	-1

---

(11) Cfr. A. M. OSTROWSKI, <u>On the convergence of the Rayleigh quotient</u> ...., II, Archive Rat. Mech. and Anal. vol. II, 1959, 427.

A. M. Ostrowski

Vari metodi per accelerare la convergenza.

Consideriamo la formula che esprime un procedimento iterativo del primo ordine :

$$\lambda_{\nu+1} = \varphi(\lambda_\nu)$$

e supponiamo che la funzione $\varphi(\lambda)$ sia derivabile con derivata continua.

Si dice che σ è un punto fisso dell'iterazione, se riesce

$$\sigma = \varphi(\sigma).$$

Distinguiamo i seguenti casi :

1) Sia $|\varphi'(\sigma)| < 1$. In tal caso, se λ_1 è abbastanza pros-
simo a σ , il procedimento iterativo è convergente e si ha :

$$(38) \qquad \lim_{\nu \to \infty} \frac{\lambda_{\nu+1} - \sigma}{\lambda_\nu - \sigma} = \varphi'(\sigma).$$

La convergenza di $\{\lambda_\nu\}$ a σ è pertanto soltanto lineare, se è $\varphi'(\sigma) \neq 0$.

2) Sia $|\varphi'(\sigma)| > 1$. La successione $\{\lambda_\nu\}$ non converge a σ
eccettuato il caso in cui uno dei λ_ν diventi uguale a σ ; più precisamente,
è possibile costruire un intorno di σ tale che, se λ_ν è contenuto in questo
intorno e non coincide con σ , i valori successivi si allontanano da σ
finchè non siano usciti dall'intorno.

3) Sia $|\varphi'(\sigma)| = 1$. E' questo un caso indeterminato, essendo
la convergenza possibile oppure no.

Se si ha convergenza, questa è molto lenta. Sussiste a proposi -

A. M. Ostrowski

to il seguente teorema :

<u>IV. Se, per $\lambda \to \sigma$, riesce</u>

$$\varphi(\lambda) - \lambda = (\lambda - \sigma)^{L+1}\left[\gamma + O(\lambda - \sigma)\right]$$

<u>e la successione $\{\lambda_\nu\}$ converge a σ, si ha</u> :

$$(39)\quad \lim_{\nu \to \infty} \quad \nu^{\frac{1}{L}}\left|\lambda_\nu - \sigma\right| = \frac{1}{\left|L\gamma\right|^{1/L}}\,. \qquad (12)$$

     Torniamo ora al caso 1), supponendo in particolare $\varphi'(\sigma) = 0$. Nell'ipotesi che la $\varphi(\lambda)$ sia di classe 2, si può affermare che

$$\lim \frac{\lambda_{\nu+1} - \sigma}{(\lambda_\nu - \sigma)^2} = \frac{1}{2}\,\varphi''(\sigma)$$

e quindi la convergenza è almeno quadratica. Nel caso opposto ci si può chiedere se è possibile ottenere da $\varphi(\lambda)$ un'altra funzione $\Phi(\lambda)$ tale che $\Phi(\sigma) = \sigma$ e $\Phi'(\sigma) = 0$. Ad esempio, conoscendo due funzioni

---

(12) Per la dimostrazione cfr. A. OSTROWSKI, <u>Sur la convergence et l'esti-mation des erreurs dans quelques procédés de résolution des équations numériques</u>, nel volume commemorativo per D. A. GRAVE, Mosca, 1940, 213-234. Una traduzione inglese di questa memoria si trova nel Techni-cal Report No. 7, 1960, degli Applied Mathematics and Statistics Labo-ratories, Stanford, University, California.

A. M. Ostrowski

di iterazione :

$$\varphi(\sigma) = \sigma, \qquad \psi(\sigma) = \sigma, \quad \text{con} \quad \varphi'(\sigma) \neq \psi'(\sigma),$$

si può **considerare la funzione**

$$(40) \qquad \Phi(\lambda) = p\,\varphi(\lambda) + (1-p)\,\psi(\lambda)$$

la **quale ha derivata nulla in** σ per $p = \dfrac{\psi'(\sigma)}{\psi'(\sigma) - \varphi'(\sigma)}$. In

particolare, scegliendo $\psi(\lambda) = \lambda$, l'espressione di $\Phi(\lambda)$ diviene:

$$\Phi(\lambda) = p\,\varphi(\lambda) + (1-p)\,\lambda$$

e la derivata si annulla nel punto σ per $p = \dfrac{1}{1 - \varphi'(\sigma)}$

Teoricamente, cioè se $\varphi'(\sigma)$ è noto, si potrebbe rendere
convergente in modo quadratico ogni procedimento iterativo dato da una fun.
zione che abbia derivata diversa da 1.

Dal **punto** di vista pratico, in generale, sarà nota solo una limi-
tazione per $\varphi'(\sigma)$. Sia $m \leq \varphi'(\sigma) \leq M$.
Sarà plausibile prendere $p = 1 / (1 - \dfrac{M+m}{2})$ e quindi

$$\Phi(\lambda) = \frac{\varphi(\lambda) - \dfrac{M+m}{2}\lambda}{1 - \dfrac{M+m}{2}} .$$

A. M. Ostrowski

Si osserva allora che

$$\left| \Phi'(\sigma) \right| = \frac{M - m}{\left| 2 - M - m \right|} < 1 \quad \text{se}$$

1) $\quad \dfrac{M + m}{2} < 1 \quad e \quad M < 1 \, ,$

2) $\quad \dfrac{M + m}{2} > 1 \quad e \quad m > 1 \, .$

A. M. Ostrowski

Lezione VI

I metodi di STEFFENSEN e HOUSEHOLDER per accelerare: la convergen= za. [13]

Supponiamo ora di conoscere tre valori successivi λ_ν, $\lambda_{\nu+1}$ $\lambda_{\nu+2}$ forniti dal procedimento iterativo $\lambda_{\nu+1} = \varphi(\lambda_\nu)$, con punto fisso σ . Formiamo il quoziente di STEFFENSEN [14] :

$$(41) \qquad \Lambda_{\nu+1} = \frac{\lambda_\nu \lambda_{\nu+2} - \lambda_{\nu+1}^2}{\lambda_\nu - 2\lambda_{\nu+1} + \lambda_{\nu+2}} \ .$$

Esprimendo $\Lambda_{\nu+1}$ per mezzo di $\Lambda_\nu = \lambda_\nu$ si ottiene

$$(42) \qquad \Lambda_{\nu+1} = \Phi(\Lambda_\nu) \ ,$$

ove si ponga

$$(43) \qquad \Phi(z) = \frac{z\,\varphi[\varphi(z)] - [\varphi(z)]^2}{z - 2\varphi(z) - \varphi[\varphi(z)]} \ .$$

Possiamo dimostrare il seguente teorema :

V. **La funzione di iterazione** $\Phi(z)$, **introdotta da STEFFENSEN, ha, nel punto fisso** σ **, derivata in modulo: minore di 1 e pertanto il procedi-**

---

(13) Cfr. A. OSTROWSKI, Uber Verfahren von Steffensen und Householder zur Konvergenzverbesserung von Iterationen, ZAMP, vol. VII, 1956, 218 - 229.

(14) Cfr. J. F. STEFFENSEN, Skand. Aktuarietidskr. 16, 1933, 64 - 72.

A. M. Ostrowski

<u>mento iterativo (42) converge almeno linearmente.</u>

Dimostrazione nell'ipotesi $\quad \varphi '(\sigma) = 1.$

Dalla (43) segue :

$$(44) \qquad \Phi(z) - z = - \frac{\left[\varphi(z) - z\right]^2}{z - 2\varphi(z) + \varphi\left[\varphi(z)\right]} \quad ;$$

si verifica allora immediatamente che l'espressione $\Phi(z) - z$ è invarian
te per un cambiamento di riferimento rappresentato da $z \rightarrow z + c$,
$\varphi(z) \rightarrow \varphi(z) + c$ e ciò consente di assumere, per semplicità, $\sigma = 0$.

La funzione $\varphi(z)$, supposta derivabile in certo numero di volte nell'intorno dell'origine, avrà lo sviluppo:

$$\varphi(z) = z + \beta z^k + \dots \qquad (\beta \neq 0, \ k \text{ intero} \geqslant 2)$$

e quindi, posto $f(z) = \varphi(z) - z$, si ha, dalla (44):

$$(45) \qquad \Phi(z) - z = - \frac{\left[f(z)\right]^2}{f\left[\varphi(z)\right] - f(z)} = - \frac{\left[f(z)\right]^2}{D(z)}$$

($D(z)$ designa brevemente il denominatore $f\left[\varphi(z)\right] - f(z)$).

In forma integrale $D(z)$ ha l'espressione :

$$(46) \qquad D(z) = \int_z^{\varphi(z)} f'(x) \, dx \quad ,$$

dalla quale, applicando il teorema della media, si deduce :

A. M. Ostrowski

$$(47) \qquad D(z) = f(z)\, f'\left[z + \vartheta\, f(z)\right] \qquad\qquad (0 < \vartheta < 1)$$

Tenendo conto delle seguenti espressioni di $f(z)$ e $f'(z)$

$$f(z) = \beta\, z^k + O(z^{k+1})\ ,$$
$$f'(z) = k\,\beta\, z^{k-1} + O(z^k)\ ,$$

si ha, dalla (47) :

$$(48) \qquad D(z) = k\,\beta^2\, z^{2k-1} + O(z^{2k})\ .$$

Sostituendo la (48) nella (44) si ricava :

$$\Phi(z) - z = -\,\frac{\beta^2\, z^{2k} + O(z^{2k+1})}{k\,\beta^2 z^{2k-1} + O(z^{2k})} = -\,\frac{z}{k} + O(z^2)$$

e quindi

$$(49) \qquad \Phi(z) = z\left(1 - \frac{1}{k}\right) + O(z^2)\ .$$

La (49) mostra che $\Phi'(\sigma) = 1 - \dfrac{1}{k} < 1$, come dovevasi dimostrare.

Si noti che il procedimento iterativo (42) riesce convergente, in base alla (49), anche se il procedimento originario non lo è.

Nal caso generale il teorema V è contenuto nel seguente teore-

A. M. Ostrowski

ma VI.

HOUSEHOLDER [15] ha proposto una generalizzazione del metodo di STEFFENSEN. Si posseggano **due** procedimenti iterativi :

$$\lambda_{\nu+1} = \varphi\,(\,\lambda_\nu\,),$$

$$\lambda_{\nu+1} = \psi\,(\,\lambda_\nu\,),$$

con lo stesso punto fisso σ e si costruisca la funzione

$$(50) \qquad \Psi_{(z)} = \frac{z\,\psi\,[\varphi\,(z)] \;-\; \varphi\,(z)\,\psi\,(z)}{z \;-\; \varphi\,(z) \;-\; \psi\,(z) \;+\; \psi\,[\varphi\,(z)]}\;;$$

sussiste il seguente teorema :

VI. <u>La iterazione</u> $\lambda_{\nu+1} = \Psi\,(\lambda_\nu\,)$, <u>nell'ipotesi che la funzione ψ abbia derivata diversa da 1 nel punto σ, risulta convergente verso σ e la convergenza è almeno quadratica.</u>

Dimostrazione.

Possiamo, senza ledere la generalità, supporre $\sigma = 0$. Nello intorno di $\sigma = 0$, si avranno per φ e ψ gli sviluppi :

$$(51) \qquad \begin{aligned} \varphi\,(z) &= \alpha\,z + \beta\,z^k + \dots \\ \psi\,(z) &= \alpha_1\,z + \beta_1\,z^{k_1} + \dots \end{aligned}$$

---

(15) Cfr. A. S. HOUSEHOLDER, <u>Principles of Numerical Analysis</u>, Mc Graw - Hill, New York 1953, 126-128.

A. M. Ostrowski

con β e β_1 diversi da zero e k e k_1 interi maggiori o uguali a 2 .

Supponiamo dapprima $\varphi'(0) = \alpha \neq 1$, $\psi'(0) = \alpha_1 \neq 1$.
Dalla (50) si deducono le espressioni :

$$(52) \qquad \Psi(z) - z = - \frac{[\varphi(z) - z] \cdot [\psi(z) - z]}{z - \varphi(z) + \psi[\varphi(z)] - \psi(z)} = - \frac{N(z)}{D(z)} ,$$

$$(53) \qquad \Psi(z) = \frac{z\, D(z) - N(z)}{D(z)} ,$$

ove si ponga

$$N(z) = [\varphi(z) - z]\,[\psi(z) - z] , \quad D(z) = z - \varphi(z) + \psi[\varphi(z)] - \psi(z).$$

Dalle (51) seguono :

$$\varphi(z) - z = (\alpha - 1) z + \beta\, z^k + \ldots, \quad \psi(z) - z = (\alpha_1 - 1)z + \beta_1\, z^{k_1} + \ldots$$

e quindi :

$$\psi[\varphi(z)] = \alpha\,\alpha_1\, z + \alpha_1\,\beta\, z^k + \alpha^{k_1}\beta_1\, z^{k_1} + O(z^{k+1} + z^{k_1+1}) .$$

Si ottengono pertanto per $D(z)$ e $N(z)$ le espressioni :

$$D(z) = (\alpha - 1)(\alpha_1 - 1) z + O(z^k + z^{k_1})$$

$$N(z) = (\alpha - 1)(\alpha_1 - 1) z^2 + (\alpha^{k_1} - 1)\,\beta_1\, z^{k_1+1} + (\alpha_1 - 1)\,\beta\, z^{k+1} +$$
$$O(z^{k+2} + z^{k_1+2})$$

e, di conseguenza, per $\Psi(z)$ vale la relazione :

A. M. Ostrowski

$$(54) \qquad \Psi(z) = \frac{z\,D(z) - N(z)}{D(z)} = \frac{O(z^{k+1} + z^{k_1+1})}{z + O(z^k + z^{k_1})} = O(z^k + z^{k_1})$$

Ne segue che, nelle ipotesi ammesse ($\alpha \neq 1$, $\alpha_1 \neq 1$), si ha per $\Psi(z)$ una convergenza almeno quadratica.

Supponiamo ora che la funzione φ abbia derivata uguale a 1 nel punto fisso dell'iterazione , $\varphi'(0) = 1$.

Posto :

$$f(z) = \varphi(z) - z \quad , \qquad g(z) = \Psi(z) - z \quad ,$$

si ha :

$$f(z) = \beta\, z^k + \dots \qquad\qquad (k \geqslant 2)$$

$$g(z) = (\,\alpha_1 - 1)\, z + \beta_1\, z^{k_1} + \dots \qquad\qquad (k_1 \geqslant 2)$$

e la (52) si scrive :

$$\Psi(z) - z = - \frac{f(z)\,g(z)}{g\,[\,\varphi(z)\,] - g(z)} = - \frac{N(z)}{D(z)} \quad .$$

Potendosi anche scrivere, applicando il teorema della media :

$$D(z) = \int_{z}^{\varphi(z)} g'(x)\,dx = f(z)\,g'\left[\,z + \vartheta\,f(z)\right] \quad , \qquad 0 < \vartheta < 1 \; ,$$

ed essendo

$$g'(z) = (\,\alpha_1 - 1) + O(z^{k_1 - 1})$$

risulta :

A. M. Ostrowski

$$\Psi(z) - z = -\frac{g(z)}{g'\left[z + \vartheta\, f(z)\right]} = -\frac{(\alpha_1 - 1)z + \beta_1 z^{k_1} + \dots}{(\alpha_1 - 1) + O(z^{k_1-1})} =$$

$$= -z + O(z^{k_1})$$

e quindi

$$(55) \qquad \Psi(z) = O(z^{k_1}) \ .$$

La (55) mostra che si ha una convergenza almeno quadratica. Il teorema è allora completamente dimostrato.

La utilità di questo risultato dipende dalla possibilità che esso offre, unito al metodo di STEFFENSEN, di ottenere una iterazione convergente in modo quadratico, partendo da una iterazione della quale non sia assicurata la convergenza. Infatti se $\varphi(z)$ è la funzione che rappresenta l'iterazione che ha σ come punto fisso, e risulta $\varphi'(\sigma) = 1$, si può ottenere con il procedimento di STEFFENSEN una funzione $\psi(z)$ per la quale $\left|\psi'(\sigma)\right| < 1$; applicando alle funzioni φ e ψ il procedimento di HOUSEHOLDER, si ottiene una iterazione con convergenza almeno quadratica.

Quanto ora detto può servire a dissipare le perplessità che potrebbe suscitare, a prima vista, la necessità di conoscere due funzioni iterative distinte (si noti che se fosse $\varphi = \psi$, la funzione Ψ verrebbe a coincidere con la funzione Φ di STEFFENSEN).

A. M. Ostrowski

Lezione VII

Altri metodi per accelerare la convergenza. Applicazione al caso dei divisori elementari non lineari.

Descriviamo ora un altro metodo per migliorare la convergenza dei procedimenti iterativi. Questo metodo è più lento degli altri, ma presenta il vantaggio di potersi applicare non solo alle forme di iterazione del $1^\wedge$ ordine (tali che $\lambda_{\nu+1} = f (\lambda_\nu))$, ma anche alle forme di iterazione di ordine infinito (tali che $\lambda_{\nu+1} = f (\lambda_1, \ldots, \lambda_\nu))$. (16)

Consideriamo allora una funzione $\varphi (x)$ definita per $x > 0$ e verificante le seguenti condizioni :

1) $\quad \varphi (x) > 0$

2) $\quad \lim_{x \to 0} \varphi(x) = 0$

3) $\quad$ se $\quad x \to o$, $y \to o$, $\quad y/x \to 1$, $\quad$ allora $\quad \dfrac{\varphi(y)}{\varphi(x)} = 1 + O(\dfrac{y}{x} - 1)$.

Una funzione $\varphi (x)$ del tipo indicato è ad esempio la funzione x^α con $\alpha > 0$.

Sussiste il seguente teorema :

---

(16) In una forma più speciale questo metodo è stato sviluppato nel libro: OSTROWSKI, <u>Solution of Equations and Systems of Equations,</u> New York, Acad. Press, 1960, Appendice I.

A. M. Ostrowski

VII. <u>Sia</u> $\{z_\nu\}$ <u>una successione convergente a</u> ζ <u>e</u> $\varphi(x)$ <u>una funzione</u> <u>del tipo suddetto. Posto</u> $\eta_\nu = |z_\nu - \zeta|$, <u>si suppone che</u>

$$\lim \; \eta_\nu / [\eta_{\nu-1} \, \varphi(\eta_{\nu-1})] = 1 \;^{(17)}.$$ <u>Introdotte le quantità positive</u>

$$\varepsilon_\nu = \left| \frac{\eta_\nu}{\eta_{\nu-1} \, \varphi(\eta_{\nu-1})} - 1 \right|, \quad \delta_\nu = |z_\nu - z_{\nu-1}|, \quad \varphi_\nu = \varphi(\delta_\nu), \quad \psi_\nu = \varphi(\eta_\nu),$$

<u>e il numero</u> $\quad \sigma_\nu = \text{segno} \, (z_{\nu+1} - \zeta) \;^{(18)}$, <u>poniamo</u>

$$\Delta_n = \text{Max} \left(\varepsilon_n, \; \varepsilon_{n+1}, \; \psi_n, \; \psi_{n-1} \right).$$

<u>Sussiste allora la seguente relazione</u> :

$$(56) \qquad \frac{Z - \zeta}{z_{n+1} - \zeta} = O(\Delta_n) ,$$

<u>con</u> $\quad Z = z_{n+1} - \dfrac{\delta_n^2 \, \varphi_n \, \sigma_n}{\delta_{n-1} \, \varphi_{n-1}} .$

Dimostrazione.

---

(17) Si osservi che per $\varphi(x) = x^\alpha$ questa formula rappresenta la convergenza di ordine $\alpha + 1$.

(18) Per segno del numero complesso z si intende il numero $\dfrac{z}{|z|} = e^{i\vartheta}$ ove ϑ denota l'argomento di z .

A. M. Ostrowski

Si ha, per ipotesi, $$\frac{\eta_\nu}{\eta_{\nu-1}\,\varphi(\eta_{\nu-1})} = 1 + O(\varepsilon_\nu)\ \text{e}$$

quindi:

$$(57) \qquad \eta_n = \eta_{n-1}\,\psi_{n-1}\,(1 + O(\varepsilon_n))\,.$$

Si ha ancora:

$$\delta_n = \left|\,(z_n - \zeta) - (z_{n+1} - \zeta)\,\right| = \left|\,z_n - \zeta\,\right|\left|1 - \frac{z_{n+1} - \zeta}{z_n - \zeta}\right|$$

e di conseguenza :

$$(58) \qquad \delta_n = \eta_n\,(1 + O(\psi_n))\,.$$

Riuscendo inoltre $\delta_n \to 0$, $\eta_n \to 0$, $\dfrac{\delta_n}{\eta_n} \to 1$, per le proprietà

della funzione φ :

$$\frac{\varphi(\delta_n)}{\varphi(\eta_n)} - 1 = O\left(\frac{\delta_n}{\eta_n} - 1\right) ,$$

e pertanto sussiste la relazione :

$$(59) \qquad \varphi_n = \psi_n\,(1 + O(\psi_n))\,.$$

Dalle (58) e (59) si trae :

A. M. Ostrowski

$$\delta_n^2 \, \varphi_n = \eta_n^2 \, \psi_n \, (1 + O(\psi_n)),$$

$$\delta_{n-1} \, \varphi_{n-1} = \eta_{n-1} \, \psi_{n-1} \, (1 + O(\psi_{n-1})),$$

e di conseguenza :

$$(60) \qquad \frac{\delta_n^2 \, \varphi_n}{\delta_{n-1} \, \varphi_{n-1}} = \frac{\eta_n^2 \, \psi_n}{\eta_{n-1} \, \psi_{n-1}} \, (1 + O(\Delta_n)).$$

Dalla ipotesi $\dfrac{\eta_\nu}{\eta_{\nu-1} \, \varphi(\eta_{\nu-1})} = 1 + O(\varepsilon_\nu)$, segue

$$\eta_n^2 = \eta_{n-1}^2 \, \psi_{n-1}^2 \, (1 + O(\varepsilon_n)).$$

Si ha allora dalla (60)

$$\frac{\sigma_n \, \delta_n^2 \, \varphi_n}{\delta_{n-1} \, \varphi_{n-1}} = \sigma_n \, \eta_{n-1} \, \psi_{n-1} \, \psi_n \, (1 + O(\Delta_n)).$$

Consideriamo ora le seguenti differenze :

$$z_{n+1} - Z = \frac{\sigma_n \, \delta_n^2 \, \varphi_n}{\delta_{n-1} \, \varphi_{n-1}} = \sigma_n \, \eta_{n-1} \, \psi_{n-1} \, \psi_n \, (1 + O(\Delta_n)),$$

A. M. Ostrowski

$$z_{n+1} - \zeta = \sigma_n \left| z_{n+1} - \zeta \right| = \sigma_n \eta_{n+1} = \sigma_n \eta_n \psi_n (1 + O(\varepsilon_{n+1})) =$$

$$= \sigma_n \eta_n \psi_n (1 + O(\Delta_n)) = \sigma_n \eta_{n-1} \psi_{n-1} \psi_n (1 + O(\Delta_n));$$

formando il quoziente :

$$\frac{z_{n+1} - Z}{z_{n+1} - \zeta} = 1 + O(\Delta_n).$$

Si ottiene la tesi.

Ritornando al metodo di STEFFENSEN e HOUSEHOLDER, vogliamo mostrare, su un esempio, come questo metodo si possa applicare anche nei casi nei quali le condizioni date nella lezione precedente non sono soddisfatte direttamente. Sia A una matrice complessa non hermitiana e consideriamo il metodo del quoziente di Rayleigh (nella sua forma originale) applicato alla matrice A , quando la successione $\{\lambda_\nu\}$ converge ad un autovalore σ di A al quale corrispondono divisori elementari con esponente massimo L maggiore di 1. Si ha allora $\lambda_{\nu+1} = \varphi(\lambda_\nu)$, ove

$$(61) \qquad \varphi(\lambda) = \frac{\eta^*(A^* - \bar{\lambda} I)^{-1} \eta}{\eta^*(A^* - \bar{\lambda} I)^{-1} (A - \lambda I)^{-1} \eta}$$

Si mostra facilmente che :

$$\varphi(\lambda) = \lambda + (\lambda - \sigma)^L \; E(\lambda),$$

A. M. Ostrowski

E (λ) essendo limitato per $\lambda \to \sigma$; ma in questo caso $\varphi(\lambda)$ è una funzione razionale delle due variabili λ e $\bar{\lambda}$ e pertanto non si può parlare di derivata nel senso abituale. E' però possibile dimostrare che la funzione E (λ) soddisfa una condizione di LIPSCHITZ, nel senso particolare che:

$$(62) \qquad E(\lambda_2) - E(\lambda_1) = O(\lambda_2 - \lambda_1),$$

quando

$$(63) \qquad \lambda_1 \to \sigma \ , \ \lambda_2 \to \sigma \ , \ \frac{\lambda_2 - \sigma}{\lambda_1 - \sigma} = 1 + O(\lambda_1 - \sigma)$$

Per mezzo di queste relazioni, l'argomentazione già applicata sopra alla funzione di STEFFENSEN $\Phi(\lambda)$, può essere modificata in modo che si abbia anche in questo caso :

$$(64) \quad \Phi(\lambda) - \sigma = (1 - 1/L) \ (\lambda - \sigma) + O(\lambda - \sigma)^2 .$$

Cambiando $\Phi(\lambda)$ e $\varphi(\lambda)$ e formando la funzione corrispondente $\Psi(\lambda)$ di HOUSEHOLDER, è possibile mostrare ancora, utilizzando le (62) e (64) che riesce :

$$(65) \qquad \Psi(\lambda) - \sigma = O(\lambda - \sigma)^2$$

cioè che l'iterazione con $\Psi(\lambda)$ converge in modo quadratico. [19]

---

(19) Cfr. A. M. OSTROWSKI, <u>On the convergence of the Rayleigh quotient</u> ....., IV, Archive Rat. Mech. and Analysis, vol. 4, 1959, 154-160, dove

A. M. Ostrowski

C'è anche un altro metodo di accelerare la convergenza che utilizza la decomposizione in prodotto infinito :

$$\frac{1}{1 - x} = 1 + x + x^2 + \ldots = \prod_{\nu=1}^{\infty} (1 - x^{2^\nu})$$

e anche le decomposizioni analoghe ove la base 2 è sostituita dal 3 .

Questo metodo è importante nel calcolo della serie di LIOUVILLE-NEUMANN e anche per la programmazione del quoziente per le macchine elettroniche [20].

./.

i calcoli conducenti alle formule (61)-(65) sono sviluppati con tutti i particolari.

(20) Cfr. A. M. OSTROWSKI, Sur une transformation de la série de LIOUVILLE-NEUMANN, C. R. Acad. des Sciences, Paris, 203, 1936, 602-605; A. M. OSTROWSKI, Sur quelques transformations de la série de LIOUVILLE-NEUMANN, C. R. , 206, 1938, 1345-1347.

A. M. Ostrowski

Lezione VIII

Il metodo del quoziente di RAYLEIGH generalizzato per i divisori elementari non lineari.

Sia $E_{\mu\nu}$ una matrice quadrata di ordine n avente l'elemento corrispondente alla riga μ-sima e alla colonna ν-sima uguale ad 1 e tutti gli altri nulli. Si ha allora

$$(66) \qquad E_{\mu\nu} = (\delta_{k\mu}\,\delta_{\nu\lambda}) \qquad\qquad (21)$$

Sussiste inoltre la seguente **regola di moltiplicazione:**

$$(67) \qquad E_{\mu\nu}\,E_{k\lambda} = \delta_{\nu k}\,E_{\mu\lambda}\;.$$

Per mezzo delle matrici $E_{\mu\nu}$ possiamo scrivere ogni matrice $A = (a_{\mu\nu})$ di ordine n nel modo seguente :

$$(68) \qquad A = \sum_{\mu,\nu}^{1,\,n} a_{\mu\nu}\,E_{\mu\nu}\;.$$

Si può allora considerare l'insieme delle matrici quadrate di ordine n come un sistema di numeri complessi con n^2 unità $E_{\mu\nu}$.

---

(21) Con $\delta_{k\mu}$ si indica il simbolo di KRONECKER.

A. M. Ostrowski

In particolare consideriamo una matrice "unità ausiliaria $\underline{U}_m$" (introdotta dall'AITKEN) di ordine m, avente la prima diagonale sopra la diagonale principale composta di elementi 1 e tutti gli altri elementi nulli. U_m può scriversi :

$$(69) \qquad U_m = \sum_{\mu=1}^{m-1} E_{\mu,\mu+1}$$

Le matrici <u>unità ausiliarie</u> godono di una proprietà molto elegante riguardante le potenze. Si ha precisamente:

$$(70) \qquad (U_m)^k = \sum_{\mu=1}^{m} E_{\mu,\mu+k} \ .$$

Dalla (70) si ottiene, in particolare :

$$(71) \qquad (U_m)^{m-1} = E_{1,m}, \quad (U_m)^m = 0, \quad (U_m)^s = 0 \quad se \quad s > m.$$

Consideriamo una matrice A_o del tipo

$$(72) \qquad A_o = \sigma I_\ell + U_\ell \qquad {}^{(22)} .$$

Sussiste il seguente

---

(22) Questa è la forma canonica di JORDAN di una matrice di ordine 1 corrispondente ad un divisore elementare di ordine 1 .

A. M. Ostrowski

LEMMA 1. <u>Sia</u> A_o <u>una matrice di tipo (72).</u> <u>Per</u> $\lambda \to \sigma$ <u>e</u> $\lambda \neq \sigma$ <u>si ha</u> :

$$(73) \qquad \frac{1}{(A_o - \lambda I)^2} = \ell \, \frac{U_\ell^{\ell-1}}{(\lambda - \sigma)^{\ell+1}} + O\left(\frac{1}{(\lambda - \sigma)^\ell}\right),$$

$$(74) \qquad \frac{U_\ell}{(A_o - \lambda I)^2} = (\ell - 1) \, \frac{U_\ell^{\ell-1}}{(\lambda - \sigma)^\ell} + O\left(\frac{1}{(\lambda - \sigma)^{\ell-1}}\right).$$

Dimostrazione :

Dalla (72) segue :

$$\frac{1}{(A_o - \lambda I)^2} = \frac{1}{(\sigma - \lambda)^2} \left[I + \frac{U_\ell}{\sigma - \lambda} \right]^{-2}.$$

D'altra parte, della decomposizione Newtoniana

$$(1 + x)^{-2} = \sum_{k=0}^{\infty} \binom{-2}{k} x^k$$

segue la seguente relazione algebrica :

$$(75) \qquad \frac{1}{(1 + x)^2} = \sum_{k=0}^{\ell-1} \binom{-2}{k} x^k + \frac{x^\ell \, P(x)}{(1 + x)^2}$$

A. M. Ostrowski

ove $\quad$ P (x) $\quad$ è un polinomio in x. Ponendo nella (75) $\quad$ $x = \dfrac{U_\ell}{\sigma - \lambda}$

si trae, in base alle (71) :

$$\left[I + \frac{U_\ell}{\sigma - \lambda} \right]^{-2} = \sum_{k=0}^{\ell-1} \binom{-2}{k} \frac{U_\ell^{\,k}}{(\sigma - \lambda)^k} \ ;$$

poichè $\quad \dbinom{-2}{k} = (-1)^k (k+1) \quad$, si ha allora:

$$(76) \qquad \frac{1}{(A_o - \lambda I)^2} = \sum_{k=0}^{\ell-1} \frac{(k+1)\, U_\ell^{\,k}}{(\lambda - \sigma)^{k+2}}$$

Considerando nella (76) il termine principale, corrispondente a $k = \ell - 1$, si deduce la (73). D'altra parte, moltiplicando i due membri della (76) per U_ℓ, si ottiene :

$$(77) \qquad \frac{U_\ell}{(A_o - \lambda I)^2} = \sum_{k=1}^{\ell} \frac{k\, U_\ell^{\,k}}{(\lambda - \sigma)^{k+1}}$$

e, prendendo il termine corrispondente a $k = \ell - 1$, si ricava la (74). Il Lemma è così completamente dimostrato.

$\qquad$ E' noto che ogni matrice A con autovalore σ è equivalente ad una matrice somma riemmanniana di matrici canoniche elementari di

A. M. Ostrowski

JORDAN; esiste quindi una matrice P non degenere tale che

$$(78) \qquad C = P^{-1} A F = \sum_{i=1}^{m} \cdot A_i + B \qquad (23) \quad ,$$

ove $A_i = \sigma I_{\ell_i} + U_{\ell_i}$ e B è una matrice avente autovalori diversi da σ . La matrice A ha l'autovalore σ di molteplicità uguale a $\sum_{i=1}^{m} \ell_i$. Sussiste il seguente

LEMMA 2. $\underline{\text{Detto}}$ $L = \max (\ell_1, \ldots, \ell_m)$ $\underline{\text{e supposto}}$ $L > 1$, $\underline{\text{per}}$ $\lambda \neq \sigma$ $\underline{\text{e}}$ $\lambda \to \sigma$ si ha :

$$(79) \qquad \frac{1}{(A - \lambda I)^2} = L \frac{H}{(\lambda - \sigma)^{L+1}} + O\left(\frac{1}{(\lambda - \sigma)^L} \right)$$

$$(80) \qquad \frac{A - \sigma I}{(A - \lambda I)^2} = (L - 1) \frac{H}{(\lambda - \sigma)^L} + O\left(\frac{1}{(\lambda - \sigma)^{L-1}} \right)$$

$\underline{\text{ove}}$ H $\underline{\text{è una matrice non nulla dipendente dalla matrice}}$ A $\underline{\text{e dalla ma-}}$

---

(23) La (78) è nota come riduzione a forma canonica di JORDAN. La e + indicano la decomposizione diagonale.

A. M. Ostrowski

<u>trice</u> P <u>che riduce</u> A <u>alla forma canonica di JORDAN</u>.

Dimostrazione.

Sia C la matrice definita dalla (78); riesce :

$$(C - \lambda I)^2 = \sum_i . (A_i - \lambda I)^2 \dotplus (B - \lambda I)^2$$

e quindi

$$(C - \lambda I)^{-2} = \sum_i . (A_i - \lambda I)^{-2} \dotplus (B - \lambda I)^{-2} .$$

Per il teorema precedente si ha :

$$(81) \quad (C - \lambda I)^{-2} = \sum_{i=1}^{m} \frac{\ell_i \, U_{\ell_i}^{\ell_i - 1}}{(\lambda - \sigma)^{\ell_i + 1}} + O\left(\frac{1}{(\lambda - \sigma)^L}\right)$$

qualora si conglobi in $O \dfrac{1}{(\lambda - \sigma)^L}$ l'addendo $(B - \lambda I)^{-2}$ che

si mantiene limitato se $\lambda \to \sigma$.

Osserviamo ora che, per la definizione di L , il termine preponderante nel secondo membro della (81) si ottiene per $\ell_i = L$. Ricordando allora che $U_{\ell_i}^s = 0$ se $s \geqslant \ell_i$, possiamo scrivere :

$$(C - \lambda I)^{-2} = \sum_{i=1}^{m} \frac{L \, U_{\ell_i}^{L-1}}{(\lambda - \sigma)^{L+1}} + O\left(\frac{1}{(\lambda - \sigma)^L}\right) .$$

Posto $H = \sum_{i=1}^{m} U_{\ell_i}^{L-1}$, riesce $H \neq 0$ e resta così dimostrata la

A. M. Ostrowski.

(79) per la C.

Si ha inoltre :

$$\frac{C - \sigma I}{(C - \lambda I)^2} = \sum_{i=1}^{n} \cdot U_{\ell_i} (A_i - \lambda I)^{-2} + (B - \sigma I)(B - \lambda I)^{-2},$$

e quindi in base al teorema precedente e all'osservazione che per $\lambda \to \sigma$ l'ultimo addendo a secondo membro si mantiene limitato, riesce :

$$\frac{C - \sigma I}{(C - \lambda I)^2} = \sum \cdot \left[(\ell_i - 1) \frac{U_{\ell_i}^{\ell_i - 1}}{(\lambda - \sigma)^{\ell_i}} + O\left(\frac{1}{(\lambda - \sigma)^{\ell_i - 1}} \right) \right].$$

Con un ragionamento perfettamente analogo al precedente si ottiene:

$$\frac{C - \sigma I}{(C - \lambda I)^2} = \sum \cdot (L - 1) \frac{U_{\ell_i}^{L-1}}{(\lambda - \sigma)^L} + O\left(\frac{1}{(\lambda - \sigma)^{L-1}} \right)$$

e quindi la (80) per la C.

Per dimostrare completamente il teorema occorre ancora far vedere che il teorema seguita a sussistere sostituendo C con A. Ciò segue dal fatto che

$$(A - \sigma I)(A - \lambda I)^{-2} = P \, (C - \sigma I)(C - \lambda I)^{-2} \, P^{-1}$$

Applichiamo alla matrice A il procedimento del quoziente di

103

A. M. Ostrowski

RAYLEIGH. Poniamo :

$$\lambda_{\nu+1} = \frac{\eta_\nu A \, \xi_\nu}{\eta_\nu \, \xi_\nu} \quad , \quad \xi_\nu = (A - \lambda_\nu I)^{-1} \alpha \, , \quad \eta_\nu = \beta (A - \lambda_\nu I)^{-1}.$$

Si ottiene allora $\lambda_{\nu+1} = \varphi(\lambda_\nu)$ con $\varphi(\lambda)$ funzione razionale di λ e più precisamente :

$$\varphi(\lambda) = \beta A (A - \lambda I)^{-2} \alpha \; / \; [\beta (A - \lambda I)^{-2} \alpha] \; .$$

La funzione $\varphi(\lambda)$ ora introdotta possiede σ come punto fisso e ammette nel σ una derivata uguale a $1 - 1/L$. Si ha infatti :

$$\varphi(\lambda) - \sigma = \frac{\beta(A - \sigma I)(A - \lambda I)^{-2} \alpha}{\beta(A - \lambda I)^{-2} \alpha}$$

e, per $\lambda \neq \sigma$ e $\lambda \to \sigma$ in base alle (79) e (80)

$$\varphi(\lambda) - \sigma = \frac{(L-1)(\lambda - \sigma)^{-L} \beta H \alpha + O(\lambda - \sigma)^{1-L}}{L(\lambda - \sigma)^{-L-1} \beta H \alpha + O(\lambda - \sigma)^{-L}} =$$

$$= \frac{(L-1)(\lambda - \sigma) \beta H \alpha + O(\lambda - \sigma)^2}{L \, \beta H \alpha + O(\lambda - \sigma)} \; .$$

Se allora si suppone $\beta H \alpha \neq 0$, si ha :

A. M. Ostrowski

$$\varphi(\lambda) - \sigma = \frac{L-1}{L}(\lambda - \sigma) + O(\lambda - \sigma)^2$$

e quindi si può enunciare il seguente teorema:

IX. <u>IL metodo del quoziente di RAYLEIGH generalizzato applicato ad una matrice dotata di divisori elementari non lineari, converge linearmente se i vettori α e β sono scelti in modo generico, cioè in modo tale che</u> $\beta H \alpha \neq 0$.

Ciò è conseguenza del fatto che $\varphi'(\sigma) = 1 - \dfrac{1}{L}$.

CENTRO INTERNAZIONALE MATEMATICO ESTIVO
(C. I. M. E.)

L. E. PAYNE

ISOPERIMETRIC INEQUALITIES FOR EIGENVALUES

AND

THEIR APPLICATIONS

ROMA - Istituto Matematico dell'Università

<u>Isoperimetric Inequalities for Eigenvalues</u>

<u>and</u>

<u>Their Applications</u>

by

L. E. Payne

University of Maryland

I. Introduction:

The classical isoperimetric inequality - - the one after which all such inequalities are named- - states that of all plane curves of given perimeter the circle encoloses the largest area. This inequality was known already to the Greeks who had also some knowledge of its analogue in three dimensions. The next isoperimetric inequality to be estabilished was perhaps that of J. Steiner (1836) who made use of an operation called symmetrization. His symmetrization gave a solid at least one plane of symmetry, preserved its volume and diminished its surface area.

In 1856 Saint-Venant [82] made the conjecture that of all cy= lindrical beams of given cross-sectional area the circulare beam has the highest torsional rigidity.

The first conjectured inequality for eigenvalues was made in 1877 by Lord Rayleigh [81]; namely, that of all membranes of given area the circle has the minimum principal frequency.

An isoperimetric inequality for electrostatic capacity was conjectured by Poincarè [66] in 1903; i. e. , of all solids of given volume the sphere has the minimum electrostatic capacity.

Around 1923, Faber [20] and Krahn [41] gave independently a proof of the Rayleigh conjecture for the membrane. (In the next section we give a slight variation of their proof.) In 1930 Szegö [94] gave the first

L. E. Payne

complete proof of the isoperimetric inequality conjectured by Poincaré. The Saint-Venant conjecture was proved by Polya [69] in 1948.

A search of the literature reveals little concentrated work in the area of isoperimetric inequalities prior to 1945. However, about that time new interest in the subject was awakened, principally by the, inves= tigations of Polya and Szegö [67, 68, 69, 95]. Their book in 1951 has given added impetus to research in this field, and during the past ten or fifteen years many new isoperimetric inequalities involving physical quan= tities from various branches of mathematical physics have been obtained [36, 38, 39, 51 - 55, 57 - 61, 70 - 75, 83 - 85, 96, 99 - 101, 106].

We restrict ourselves in this paper to a consideration of iso= perimetric inequalities for eigenvalues. In fact we concern ourselves primarily with those arising in membrane and plate theory. Even then we are able to cover only a few of the most interesting and important isoperi= metric inequalities.

In the next section we define the various eigenvalues problems which are to be considered. In sections III through VI we prove a number of interesting isoperimetric inequalities. In sections VI and VII, we present two useful applications of eigenvalue inequalities. Finally, we demonstrate in section VIII the use of isoperimetric inequalities in deri= ving "optimal" a priori bounds for solutions of boundary value problems.

We shall not discuss in this paper the numerous eigenvalue inequalities in one dimensional problems-those of vibrating strings, vi= brating rods, etc. - - which have been derived in the literature (see e.g., [3, 7 - 9, 42, 50, 86, 88]). Neither shall we present the various isoperimetric inequalities relating the eigenvalues of (A) - (E) to those

L. E. Payne

of analogous finite difference problems $\left[\,24,\ 28,\ 36,\ 71,\ 100,\ 101.\right]$. The reader will undoubtelly be aware of various other isoperimetric inequalities which are not mentioned in this paper, as well as other applications for those which are mentioned. This paper is not meant as an axhau̲ stive survey, but rather as an introduction to a number of fascinating isoperimetric inequalities for eigenvalues, together with a few interesting applications.

II. Problems to be Considered:

We consider the following eigenvalue problem defined on a bounded N-dimensional region R_N with boundary C_N:

$$(A) \qquad \Delta u + \lambda u = 0 \qquad \text{in } R_N$$
$$u = 0 \qquad \text{on } C_N$$

$$(B) \qquad \Delta v + \mu v = 0 \qquad \text{in } R_N$$
$$\partial v / \partial n = 0 \qquad \text{on } C_N$$

$$(C) \qquad \Delta^2 W + \Lambda W = 0 \qquad \text{in } R_N$$
$$W, \frac{\partial W}{\partial n} = 0 \qquad \text{on } C_N$$

$$(D) \qquad \Delta^2 \chi - \Omega \chi = 0 \qquad \text{in } R_N$$
$$\chi, \partial \chi / \partial n = 0 \qquad \text{on } C_N$$

$$(E) \qquad \Delta h = 0 \qquad \text{in } R_N$$
$$\frac{\partial h}{\partial n} - q h = 0 \qquad \text{on } C_N .$$

In the above problems Δ denotes the Laplace operator and $\partial / \partial n$ denotes the normal derivative on C_N . The constants λ , μ ,

111

L. E. Payne

Λ, Ω, and q are the eigenvalues of the various problems. Because of the forms of the operators it is clear that all eigenvalues are real and non-negative.

In two dimensions the eigenvalues λ_i of (A) and the eigen= values μ_i of (B) may be interpreted as the squares of the vibration frequencies of a fixed and of a free membrane. In three dimensions both the λ_i and μ_i have interpretations in acoustics, diffusions theory, electromagnetic theory , etc. In two dimensions the eigenvalues Λ_i of (C) and Ω_i of (D) arise in the theory of elastic plates. The quantities Λ_i are proportional to the critical buckling loads for a clamped elastic plate occupying R_2, while the Ω_i are proportional to the squares of the vibration frequencies of a clamped elastic plate defined over R_2. The eigenvalues q of problem (E) are the so-called Stekloff eigenva= lues [91].

Most of the inequalities which we consider in this paper will involve the first non-zero eigenvalues of problems (A) - (E). Regardless of N we shall refer to the problems throughout as membrane, plate or Stekloff problems.

We order the eigenvalues λ_i in the following way

$$(2:1) \qquad \lambda_1 \leqslant \lambda_2 \leqslant \lambda_3 \leqslant \quad \ldots.$$

The other eigenvalues are similarly ordered.

Throughout this paper we assume that the bounding surface C is sufficiently smooth so that all of the applications of the divergence theorem remain valid.

L. E. Payne

III. The Faber-Krahn Inequality and Its Extensions :

We give in this section a slight variation of the Faber-Krahn proof of the Rayleigh conjecture in N-dimensions (Krahn actually proved the result in N-dimensions) and indicate a way of extending their result. We prove then first the following theorem:

__Theorem I.__ __The first eigenvalue in the fixed membrane problem for R is not smaller than that for the sphere of the same N-volume.__

It is well known (see e. g., $[15]$) that the eigenfunction u_1, corresponding to the first eigenvalue λ_1, of (A) does not change sign R_N. If the first eigenfunction is taken to be positive, then clearly u_1, can have no relative minimum in R_N. For convenience we now drop the subscript on u_1.

Let $R_N(\bar{u})$ (with boundary $C_N(\bar{u})$) denote that portion of R_N over which $u \geqslant \bar{u}$. At each point on the surface $u = \bar{u}$, a curvilinear coordinate system is introduced, which involves any convenient (N-1) - dimensional coordinate system in the surface $u = \bar{u}$, and the normal to the surface. 1)

---

1) By an easy generalization of results of O. D. Kellogg (__Foundations of Potential Theory__, Dover Publications, Inc. (1953) pp. 273-277) it can be shown that the introduction of this system of coordinates is legitimate. Another proof of the Rayleigh conjecture by L. Tonelli (__Sur un problème de Lord Rayleigh__, Monatsh. Math. Phys., Vol. 37 (1930) pp. 253-280) avoids the introduction of such a system of coordinates.

L. E. Payne

For any function F defined in R_n it then follows that

$$(3.1) \qquad \int_{R_N} F \, dv = \int_0^{u_n} \int_{C_N(\bar{u})} \frac{F}{|\operatorname{grad} \bar{u}|} \, ds \, du$$

(see e.g. Polya and Szegö [76]). Here u_n denotes the maximum value of u in R_N . We note then that λ_1 , satisfies

$$(3.2) \qquad \lambda_1 = \frac{\int_{R_N} |\operatorname{grad} u|^2 \, dv}{\int_{R_N} u^2 \, dv} = \frac{\int_0^{u_n} \oint_{C_N(\bar{u})} |\operatorname{grad} \bar{u}| \, ds \, d\bar{u}}{\int_0^{u_n} \oint_{C_N(\bar{u})} \frac{\bar{u}^2}{|\operatorname{grad} \bar{u}|} \, ds \, d\bar{u}} \ .$$

We introduce the notation

$$(3.3) \qquad \Psi_N(u) = \int_{\tilde{u}}^{u_n} \oint_{C_N(\bar{u})} \frac{ds}{|\operatorname{grad} \bar{u}|} \, d\bar{u} \ .$$

Clearly $\Psi_N(\tilde{u})$ is the volume of $R_N(\tilde{u})$. Then from (3.3) we have

$$(3.4) \qquad -\frac{d \Psi_N(\bar{u})}{d\bar{u}} = \oint_{C_N(\bar{u})} \frac{ds}{|\operatorname{grad} \bar{u}|} \ .$$

Hence by Schwarz's inequality

114

L. E. Payne

$$(3.5) \qquad - \frac{d\,\Psi_N(\bar{u})}{d\bar{u}} \oint_{C_N(\bar{u})} \left| \operatorname{grad} u \right| ds \;\geqslant\; S_N^2(\bar{u}) \,.$$

where $S_N(\bar{u})$ is the surface area of $C_N(\bar{u})$. By the classical isoper-imetric inequality we have

$$(3.6) \qquad S_N^2(\bar{u}) \;\geqslant\; \left[\frac{N\,\Psi_N(\bar{u})}{\omega_N} \right]^{2\left(\frac{N-1}{N}\right)} \omega_N^{\,2} \,.$$

Here ω_N denotes the surface area of the unit sphere in N-dimensions. Inserting (3.6) into (3.5) we obtain

$$(3.7) \qquad - \frac{d\,\Psi_N(\bar{u})}{d\bar{u}} \oint_{C_N(u)} \left| \operatorname{grad} u \right| ds \;\geqslant\; \left[\frac{N\,\Psi_N(\bar{u})}{\omega_N} \right]^{2\left(\frac{N-1}{N}\right)} \omega_N^{\,2} \,.$$

Thus

$$(3.8) \qquad \int_{R_N} \left| \operatorname{grad} u \right|^2 dv \;\geqslant\; N^{\,2\left(\frac{N-1}{N}\right)} \omega_N^{\,2/N} \int_{\tilde{u}}^{u_n} \left[\Psi_N(\bar{u}) \right]^{2\left(\frac{N-1}{N}\right)} \left(\frac{-d\bar{u}}{d\Psi_N(\bar{u})} \right) d\bar{u} =$$

$$= \omega_N^{\,2/N}\, N^{\,2\left(\frac{N-1}{N}\right)} \int_{o}^{V_N} \left[\Psi_N \right]^{2\left(\frac{N-1}{N}\right)} \left[\frac{du}{d\Psi_N} \right]^2 d\Psi_N \,,$$

where V_N is the volume of R_N. Clearly since $u(V_N) = 0$

L. E. Payne

we have that

$$(3.9) \qquad \int_0^{V_N} \psi_N^{2\left(\frac{N-1}{N}\right)} \left[\frac{du}{d\psi_N}\right]^2 d\psi_N \geq \lambda^2 \int_0^{V_N} u^2 \, d\psi_N,$$

where λ is the lowest eigenvalue of the problem

$$(3.10) \qquad \frac{d}{d\psi}\left[\psi^{2\left(\frac{N-1}{N}\right)}\frac{dP}{d\psi}\right] + \lambda^2 P = 0, \qquad 0 \leq \psi \leq V_N,$$

$$P(V_N) = 0, \qquad \lim_{\psi \to 0} \psi^{2\left(\frac{N-1}{N}\right)}\frac{dP}{d\psi} = 0$$

The solution to this problem si given by

$$(3.11) \qquad P = \psi^{-\left(\frac{N-2}{2N}\right)} J_{\frac{N-2}{2}}\left[N\lambda\psi^{1/N}\right]$$

where λ is to be determined so that $P(V_N) = 0$.
That is

$$(3.12) \qquad \lambda = j_{\frac{N-2}{2}} \Big/ N V_N^{1/N}$$

where $j_{\frac{N-2}{2}}$ is the first zero of the Bessel function $J_{\frac{N-2}{2}}$

L. E. Payne

Combining (3.9) and (3.12), and inserting the result into (3.2), we obtain finally

$$(3.13) \qquad \lambda_1 \geq \frac{\omega_N^{2/N} \, N^{2\left(\frac{N-1}{N}\right)} \int_0^{V_N} \psi_N^{2\left(\frac{N-1}{N}\right)} \left[\frac{du}{d\psi_N}\right]^2 d\psi_N}{\int_0^{V_N} u^2 \, d\psi_N} \geq$$

$$\geq \left[\frac{\omega_N}{N V_N}\right]^{2/N} j^2 \frac{N-2}{2} \, ,$$

Since the equality sign holds if R_N is an N-sphere this is the desired Faber-Krahn inequality and the theorem is proved. In fact it follows easily from our proof that the equality sign is valid in (3.13) if and only if R_N is an N-sphere.

We seek now some possible extensions of the Faber-Krahn inequality in two dimensions. To this end we introduce a cartesian coordinate system in such way that R_N lies in the half space $x_i > 0$. Then u_i is representable as

$$(3.14) \qquad u_i = x_i \, \varphi$$

where $\varphi \in C$ in the closure of R_N and vanishes on C_2', the portion (or portions) of C_2 in the open region $x_i > 0$. Clearly φ satisfies

L. E. Payne

$$(3.15) \qquad \Delta\varphi + \frac{2}{x_1}\frac{\partial\varphi}{\partial x_1} + \lambda_1 \varphi = 0 \text{ in } R_2 , \qquad \varphi = 0 \text{ on } C_2' .$$

The function φ may be interpreted as the first eigenfunction in the fixed membrane problem for a 4-dimensional body symmetric about the x_1-axis (see Weinstein $[105]$). By the Faber-Krahn inequality in 4-dimensions we have then

$$(3.16) \qquad \lambda_1 \geqslant \left[\frac{\omega_4}{4 V_4}\right]^{1/2} j_1^2 .$$

But V_4 is proportional to I_1, the moment of inertia (unit density) of V_2 about the x_2-axis, i.e.,

$$(3.17) \qquad I_1 = \int_{R_2} x_1^2 \, dA.$$

In fact (3.16) may be written as

$$(3.18) \qquad \lambda_1 \geqslant 1/2 \left[\pi/2 I_1\right]^{1/2} j_1^2 .$$

Equality clearly holds if R_2 is a semicircle.

Similarly, if the coordinate system is chosen in such a way that R_2 lies in the quadrant $x_1 > 0$, $x_2 > 0$, and u_1 is represented as

$$(3.19) \qquad u_1 = x_1 x_2 \Phi$$

L. E. Payne

then $\bar{\Phi}$ must satisfy

$$(3.20) \qquad \Delta \bar{\Phi} + \frac{2}{x_1} \frac{\partial \bar{\Phi}}{\partial x_1} + \frac{2}{x_2} \frac{\partial \bar{\Phi}}{\partial x_2} + \lambda_1 \bar{\Phi} = 0 \text{ in } R_2, \quad \bar{\Phi} = 0 \text{ on } C_2''$$

where C_2'' is the portion of C_2 for which $x_1 > 0$ and $x_2 > 0$.

Again $\bar{\Phi}$ may be interpreted as an eigenfunctio of a 6-dimensional membrane with both the x_1 and x_2 axes as axes of symmetry. The volume of this body is proportional to J_{12} where

$$(3.21) \qquad J_{12} = \int_{R_2} x_1^2 \, x_2^2 \, d\dot{V}.$$

The Faber-Krahn inequality in 6-dimensions then yields the inequality

$$(3.22) \qquad \lambda_1 \geqslant 1/2 \left[\frac{\pi}{12 \, J_{12}} \right]^{1/3} j_2^2 .$$

Finally, let us suppose that R_2 lies interior to the wedge bounded by $0 \leqslant \theta \leqslant \pi/n$ where n is an integer. In this case we set

$$(3.23) \qquad u = r^n \sin n\theta \, \psi \, , \qquad r^2 = x_1^2 + x_2^2$$

The function ψ satisfies the equation

L. E. Payne

$$(3.24) \quad \frac{\partial^2 \psi}{\partial r^2} + \frac{(2n-1)}{r} \frac{\partial \psi}{\partial r} + \frac{1}{r^2} \frac{\partial^2 \psi}{\partial \theta^2} + \frac{2n}{r^2} \cot n\theta \frac{\partial \psi}{\partial \theta} + \lambda_1 \psi =$$

$$= 0 \text{ in } R_2 , \qquad \psi = 0 \text{ on } C_2^* .$$

In this case C^* is the portion of C which lies in the open region $0 < \theta < \pi/n$. From Krahn's result $\begin{bmatrix} 41 \end{bmatrix}$ in $2(n+1)$ dimensions it follows as before that

$$(3.25) \qquad \lambda_1 \geq \left[\frac{\pi}{4} \frac{K_n}{n(n+1)} \right]^{-1^{\frac{1}{n+1}}} j_n^2 .$$

where

$$(3.26) \qquad K_n = \left[\int_{R_2} r^{2n} \sin^2 n\theta \, dA \right] .$$

With such results for integral values of n, it is natural to conjecture the following theorem.

Theorem II: If R_2 lies interior to the wedge $0 \leq \theta \leq \pi/\alpha$ for any real $\alpha \geq 1$ then

$$(3.27) \qquad \lambda_1 \geq \left[\frac{\pi}{4} \frac{K_\alpha}{\alpha(\alpha+1)} \right]^{-1^{\frac{1}{\alpha+1}}} j_\alpha^2 ,$$

where

$$(3.28) \qquad K_\alpha = \int_{R_2} r^{2\alpha} \sin^2 \alpha\theta \, dA ,$$

L. E. Payne

with equality if and only if R_2 is a circular sector.

This result has recently been proved by Payne and Weinberger $\begin{bmatrix} 58 \end{bmatrix}$.

We now apply Theorem II to a simple example and note that there appears to be no obvious systematic method of determining the origin to give the best lower bounds.

Let G be a right isoceles triangle with equal sides of unit length. If the origin is taken at the midpoint of the hypotanuse we have $\alpha = 1$ and hence

$$(3.29) \qquad \lambda_1 \geqslant 6.714.$$

If one of the acute-angled corners is chosen as the origin we have $\alpha = 4$. Then we obtain

$$(3.30) \qquad \lambda_1 \geqslant 6.775.$$

Finally, if the right-angled corner is taken as origin we have $\alpha = 2$ which gives

$$(3.31) \qquad \lambda_1 \geqslant 6.902.$$

All of these bounds are better than the lower bound 6.029 obtained from the Faber-Krahn inequality. The exact value is known to be 7.025.

The results for the cases in which the body lies in a half plane or in a quadrant are easily extended to higher dimensions. It appears likely that analogues in N-dimensions of the results for various wedge angles can be established, although this has not been carried out.

L. E. Payne

IV. An Optimal Poincaré Inequality for Convex Domains.

The Poincare constant K is any upper bound for μ_2^{-1} where μ_2 is the first non-zero eigenvalue of (B). The existence of K for quite general regions is well-known (see e.g., [15, 65]); however, explicit values for K (or equivalently lower bounds for μ_2) are in general not known. In this section we prove the following theorem first established by Payne and Weinberger [59] .

Theorem III.   The first non-zero eigenvalue μ_2 in the free membrane problem for a convex region R_N satisfies the inequality

$$(4.1) \qquad \mu_2 \geqslant \pi^2 D^{-2}$$

where D is the diameter of R_N .

The inequality (4.1) is the best bound which can be given in terms of the diameter alone in the sense that $\mu_2 D^2$ tends to π^2 for a parallelepiped all but one of whose dimensions shrink to zero.

We shall prove the result only in two dimensions; however, the same method applied in the appropriate manner will yield the same result for any N.

The quantity μ_2 is defined as the infimum of the quotient

$$(4.2) \qquad \int_{R_2} |\text{grad } \varphi|^2 \, dA \Big/ \int_{R_2} \varphi^2 \, dA$$

among functions which have bounded second derivatives in R and satisfy $\int_{R_2} \varphi \, dA = 0$. Now for any such function φ we consider

L. E. Payne

the set of lines which divide R_2 into subregions of equal area. It follows by continuity that at least one line divides R_2 into two convex subdomains of equal area over each of which the integral of φ vanishes. We now divide each subdomain into two additional convex subdomains of equal area over each of which the integral of φ vanishes.

By cintinuing this process we arrive after a finite number of steps at a division of R_2 into convex subdomains R_2^{ν} of arbitrarily small area A_{ν}. Over each of the R_2^{ν} the integral of φ vanishes.

We introduce a rectangular coordinate system in any R_2^{ν} in the following way :

Let the x_2 -axis lie along the line of maximum diameter of R_2^{ν} and take x_1 -axis perpendicular to the x_2 -axis and tangent to one end of R_2^{ν}. Let L_{ν} be the length of the projecton of R_2^{ν} on the x_2 -axis. Clearly $L_{\nu} \lessgtr D$. Let $p(y)$ denote the length of intersection of R_2^{ν} with the line $x_2 = y$. Since the area of R_2^{ν} can be made arbitrarily small, then after a sufficiently large number of subdivisions $p(y)$ may be made to satisfy

$$(4.3) \qquad\qquad p(y) \leqslant \varepsilon$$

for any prescribed ε . Because of the convexity of R_2^{ν}, $p(y)$ must be a convex function of y.

Let M be a uniform bound for φ , its first derivative in R_2. (Actually M could denote a uniform bound for φ its first derivative and the integral $\int_0^{\varepsilon} |\varphi_{\xi\,y}| \, d\xi$ if $\varphi \in c^2$ in the closure of R_2) Then by the mean value theorem

L. E. Payne

$$(4.4) \quad \left| \int_{R_2^\vee} \left(\frac{\partial \varphi}{\partial x_2}\right)^2 dA - \int_0^{L_\vee} p(y) \left[\varphi'(o,y)\right]^2 dy \right| \leq 2 M^2 A_\vee \varepsilon .$$

$$(4.5) \quad \left| \int_{R_2^\vee} \varphi^2 dA - \int_0^{L_\vee} p(y)\left[\varphi(o,y)\right]^2 dy \right| \leq 2 M^2 A_\vee \varepsilon ,$$

$$(4.6) \quad \left| \int_{R_2^\vee} \varphi \, dA - \int_0^{L_\vee} p(y) \, \varphi(o,y) \, dy \right| \leq M A_\vee \varepsilon . \quad \text{Thus}$$

$$(4.7) \quad \int_{R_2^\vee} |\text{grad } \varphi|^2 \, dA \geq \int_{R_2^\vee} \left(\frac{\partial \varphi}{\partial x_2}\right)^2 dA \geq \int_0^{L} p(y) \left[\varphi'(o,y)\right]^2 dy -$$

$$- 2 M^2 A_\vee \varepsilon ,$$

Also

$$(4.8) \quad \int_{R_2^\vee} \varphi^2 \, dA \leq \int_0^{L_\vee} p(y) \left[\varphi(o,y)\right]^2 dy + 2 M^2 A_\vee \varepsilon$$

and

$$(4.9) \quad \int_{R_2^\vee} p(y) \, \varphi(o,y) \, dy \leq M A_\vee \varepsilon .$$

We require now the following inequality

$$(4.10) \quad \int_0^{L} p(y) \left[\varphi'(o,y)\right]^2 dy \geq \pi^2 L_\vee^{-2} \left\{ \int_0^{L} p(y) \left[\varphi(o,y)\right]^2 dy - A_\vee^{-1} \left(\int_0^{L} p(y) \, \varphi(o,y) \, dy \right)^2 \right\} .$$

L. E. Payne

However before we prove this inequality let us demonstrate that knowledge of such an inequality would lead to the desired result (4.1).   Note that (4.10) together with (4.7) - (4.9) would yield

$$(4.11) \qquad \int_{R_2^v} \left| \text{grad } \varphi \right|^2 dA \geqslant \pi^2 D^{-2} \left\{ \int_{R_2^v} \varphi^2 \, dA - M^2 A_v \, \varepsilon \, (2+\varepsilon) \right\} - 2 \, M^2 \, A_v \, \varepsilon$$

or

$$(4.12) \qquad \int_{R_2^v} \left| \text{grad } \varphi \right|^2 dA \geqslant \pi^2 D^{-2} \left\{ \int_{R_2^v} \varphi^2 \, dA \right\} - M^2 A_v \, \varepsilon \left[2 + \pi^2 D^{-2} (2+\varepsilon) \right] .$$

A summation of the inequalities (4.12) would now be made over all subdomains R_2^v .   The sum of A_v is the area of R_2^v . Since ε is arbitrarily small we would then obtain the inequality

$$(4.13) \qquad \int_{R_2} \left| \text{grad } \varphi \right|^2 dA \geqslant \pi^2 D^{-2} \int_{R_2} \varphi^2 \, dA .$$

Since φ is any function with bounded second derivatives satisfying $\int_{R_2} \varphi \, dA = 0$, we would have from (4.2)

$$(4.14) \qquad \mu_2 \geqslant \pi^2 D^{-2} .$$

L. E. Payne

Thus, once the one dimensional inequality (4.10) is established Theorem II is proved.

We prove then the following Lemma:

<u>Lemma I. Let $p(y)$ be a non-negative convex function of y defined on the interval $o \leq y \leq L$; then for any piecewise continuously differentiable function $\varphi(y)$ satisfying</u>

$$(4.15) \qquad \int_0^L p(y)\, \varphi(y)\, dy = 0$$

<u>it follows that</u>

$$(4.16) \qquad \int_0^L p(y) \left[\varphi'(y) \right]^2 dy \geq \pi^2 L^{-2} \int_0^L p(y) \left[\varphi(y) \right]^2 dy.$$

Assume for the moment that p is strictly positive and twice differentiable. Then the function ψ which minimizes the quotient

$$(4.17) \qquad \int_0^L p(y) \left[\psi'(y) \right]^2 dy \Big/ \int_0^L p(y) \left[\varphi(y) \right]^2 dy$$

among the functions satisfying (4.15) must satisfy

$$(4.18) \qquad \left[p\, \psi' \right]' + V\, p\, \psi = 0, \quad \psi'(y) \neq 0, \quad 0 < y < L,$$

$$\psi'(o) = \psi'(L) = 0$$

L. E. Payne

where V is the minimum value of the quotient (4.17). We now introduce the new variable

$$(4.19) \qquad W = \psi' \, p^{1/2} .$$

The function W satisfies the Sturm-Liouville system

$$(4.20) \qquad W'' + \left[\frac{1}{2} \frac{p''}{p} - \frac{3}{4} \left(\frac{p'}{p} \right)^2 \right] W + V W = 0 \; ; \quad 0 < y < L$$

$$W(o) = W(L) = 0 .$$

Because of the convexity of p, the term in square brackets is non-positive. Hence, multiplying (4.20) by W and integrating by parts we find

$$(4.21) \qquad V \geq \frac{\displaystyle\int_o^L \left[W' \right]^2 dy}{\displaystyle\int_o^L W^2 \, dy} .$$

Since $W(o) = W(1) = 0$, the quotient is bounded below by the first eigenvalue of the vibrating string fixed at its ends, i.e.,

$$(4.22) \qquad V \geq \pi^2 / L$$

Thus the lemma is proved if $p > o$ and twice differentiable in y.

Now if $\widetilde{\varphi}$ (y) is any function defined on the interval $o \leq y \leq L$, the function

L. E. Payne

$$(4.23) \qquad \varphi(y) = \tilde{\varphi}(y) - \left[\int_0^L p\, dy\right]^{-1} \int_0^L p\, \tilde{\varphi}\, dy$$

will satisfy (4.15). Hence for $p(\, \xi\, c^2) > 0$, (4.16) implies that

$$(4.24) \qquad \int_0^L p\left[\tilde{\varphi}'\right]^2 dy \geq \pi^2 L^{-2}\left\{\int_0^L p\, \tilde{\varphi}^{\,2}\, dy - \right.$$

$$\left. - \left[\int_0^L p\, dy\right]^{-1}\left[\int_0^L p\, \tilde{\varphi}\, dy\right]^2\right\}.$$

Clearly (4.24) holds for the uniform limit of admissable functions p. Hence (4.24) is valid for any non-negative convex function of y. This proves lemma I and thus establishes the desired inequality (4.16).

We were interested in this section only in the lemma that was required for proving theorem II. The proof, however, actually shows that if $p^{-\frac{1}{2}}$ is a concave function of y the eigenvalues V_h of (4.18) satisfy the inequality

$$(4.25) \qquad V_h \geq (h-1)^2\, \pi^2\, L^{-2}$$

with equality if and only if $p^{-\frac{1}{2}}$ is linear in y.

L. E. Payne

V. Additional Eigenvalue Inequalities :

In this section we prove a conjecture of Weinstein [103] which relates certain membrane eigenvalues to those of the buckling problem for a clamped plate of the same shape. The proof of this conjecture was given by Payne [51]. Other interesting inequalities, both proved and conjectured are discussed in the latter part of the section.

We prove first the Weinstein conjecture :

Theorem IV. The first eigenvalues in the buckling problem for a clamped plate is not less than the second eigenvalues of the membrane of the same shape which is fixed on the boundary, i. e.,

$$(5.1) \qquad \Lambda_1 \geq \lambda_2$$

This inequality relates eigenvalues of problems (A) and (C) defined on the same domain.

To prove the theorem we make use of the following minimum principle for λ_2 :

$$(5.2) \qquad \lambda_2 = \min \frac{\int_{R_N} |\operatorname{grad} \psi|^2 \, dv}{\int_{R_N} \psi^2 \, dv}$$

where ψ is any sufficiently smooth function defined in R_N which vanishes on C_N and satisfies the condition

$$(5.3) \qquad \int_{R_N} u_1 \, \psi \, dv = 0 .$$

L. E. Payne

Let trial functions ψ_α be defined as follows:

$$(5.4) \qquad \psi_\alpha = a_\alpha w_1 + \frac{\partial w_1}{\partial x_\alpha} \, , \qquad \alpha = 1, 2, \ldots, N.$$

where w_1 is the first eigenvalue of problem (C) defined on R_N, and the constants a_α are so chosen that (5.3) is satisfied. We may assume that $\int_{R_N} u_1 w_1 \, dv$ is not zero since if it were we would have by (5.2)

$$(5.5) \quad \lambda_2 \; \frac{\int_{R_N} |\operatorname{grad} w_1|^2 dv}{\int_{R_N} w_1^2 \, dv} \; \leqslant \; \frac{\int_{R_N} (\Delta w_1)^2 dv}{\int_{R_N} |\operatorname{grad} w_1|^2 \, dv} \; = \; \Lambda_1$$

and the theorem would be established. The second inequality results from an application of the Schwarz inequality to the identity

$$(5.6) \qquad \int_{R_N} |\operatorname{grad} w_1|^2 \, dv = - \int_{R_N} w_1 \, \Delta \, w_1 \, dv.$$

In view of the boundary conditions satisfied by w_1, it follows from the divergence theorem that

$$(5.7) \qquad \int_{R_N} w_1 \frac{\partial w_1}{\partial x} \, dv = 0 \; ; \; \sum_{i=1}^{N} \int_{R_N} \frac{\partial^2 w}{\partial x_i \partial x_\alpha} \frac{\partial w}{\partial x_i} \, dv = 0.$$

L. E. Payne

Thus from the minimum principle (5.2),

$$(5.8) \qquad \lambda_2 \leq \frac{a_\alpha^2 \int_{R_N} |\text{grad } w_1|^2 \, dv + \int_{R_N} \frac{\partial^2 w_1}{\partial x_i \partial x_\alpha} \frac{\partial^2 w_1}{\partial x_i \partial x_\alpha} \, dv}{a_\alpha^2 \int_{R_N} w_1^2 \, dv + \int_{R_N} \frac{\partial w_1}{\partial x_\alpha} \frac{\partial w_1}{\partial x_\alpha} \, dv}$$

where summation is to be carried out over Latin subscripts but not over

Greek subscripts. Adding the inequalities we obtain

$$(5.9) \qquad \lambda_2 \leq \frac{\sum_{j=1}^{N} a_j^2 \int_{R_N} |\text{grad } w_1|^2 \, dv + \sum_{i,j=1}^{N} \int_{R_N} \frac{\partial^2 w_1}{\partial x_i \partial x_j} \frac{\partial^2 w_1}{\partial x_i \partial x_j} \, dv}{\sum_{j=1}^{N} a_j^2 \int_{R_N} w_1^2 \, dv + \int_{R_N} |\text{grad } w_1|^2 \, dv} \, .$$

But

$$(5.10) \qquad \frac{\sum_{i,j=1}^{N} \int_{R_N} \frac{\partial^2 w_1}{\partial x_i \partial x_j} \frac{\partial^2 w_1}{\partial x_i \partial x_j} \, dv}{\int_{R_N} |\text{grad } w_1|^2 \, dv} = \frac{\int_{R_N} |\Delta w_1|^2 \, dv}{\int_{R_N} |\text{grad } w_1|^2 \, dv} = \Lambda_2 \, .$$

Thus, in view of (5.5) it follows that

$$(5.11) \qquad\qquad \lambda_2 \leq \Lambda_1$$

and the theorem is proved. It can be shown without too much difficulty

that the only region for which the equality sign holds, is the sphere (see

[51]   ).

Polya and Szegö [76]   have shown that

L. E. Payne

$$(5.12) \qquad \Omega_1 > \Lambda_1 \lambda_1 \ .$$

This leads then to the inequality

$$(5.13) \qquad \Omega_1 > \lambda_1 \lambda_2 \ .$$

Payne $\begin{bmatrix} 54 \end{bmatrix}$ has shown further that for R_N convex

$$(5.14) \qquad \Lambda_1 \leq 4 \lambda_1 \ .$$

This inequality is the best possible in the sense that the equality sign holds in the limit for a strip region. He showed also that for convex R_N

$$(5.15) \qquad \Omega_1 < \frac{16}{3} \lambda_1^2$$

and

$$(5.16) \qquad 1/2 \, \Lambda_1 < \Omega_1^{1/2} < \Lambda_1 \ .$$

Before passing on to other isoperimetric inequalities, we mention a few conjectured inequalities involving membrane and plate eigenvalues. We consider only the case $N = 2$.

Conjecture I:

The first eigenvalue Λ_1 in the vibration problem for a clamped plate is not less than that for the circle of the same area.

Conjecture II:

The first eigenvalue Ω_1 in the buckling problem for a clamped

L. E. Payne

p'late is not less than that for the circle of the same area.

Conjecture III:

The nth eigenvalue λ_n in the fixed membrane problem for R_2 satisfies the inequality

$$(5.17) \qquad \lambda_n \geqslant 4 \pi n A^{-1}$$

where A is the area of R_2.

Conjecture IV:

The nth eigenvalue μ_n in the free membrane problem for R_2 satisfies the inequality

$$(5.18) \qquad \mu_n \leqslant 4 \pi (n - 1) A^{-1} .$$

Conjecture I was made by Lord Rayleigh [81] ; conjecture II is due to Polya and Szegö [76] , and conjectures III and IV are due to Poyla [74] . The first two conjectures were proved [76] under the hypothesis that in each case the first eigenfunction does not change sign in R. But under what conditions this hypothesis is satisfied, has not been determined. Conjectures III and IV were proved by Polya [74] for special types of "space filling" domains.

We list also two other interesting isoperimetric inequalities which in a certain way complement the Faber-Krahn inequality. We state the theorems without proof, referring the reader to the original papers.

L. E. Payne

Theorem V: The first eigenvalue λ_1 in the fixed membrane problem for a simply connected region R_2 is not greater than that for the annular domain (concentric circular boundaries) of the same area fixed on the outer boundary, whose perimeter is equal to that of C, and free along the inner boundary.

Theorem VI: The first non-zero eigenvalue μ_2 in the free membrane problem for R_N is not greater than that for the sphere of the same volume.

Theorem V is an extension of earlier results of Makai $\begin{bmatrix} 44 \end{bmatrix}$, $\begin{bmatrix} 45 \end{bmatrix}$, and Polya $\begin{bmatrix} 75 \end{bmatrix}$. It was proved by Payne and Weinberger $\begin{bmatrix} 60 \end{bmatrix}$. Inequalities of a similar nature for membranes defined in a multiply connected region have been obtained by J. Hersch (unpublished).

Theorem VI was formulated in two-dimensions by Korhauser and Stakgold $\begin{bmatrix} 40 \end{bmatrix}$ and proved for simply connected regions by Szegö $\begin{bmatrix} 96 \end{bmatrix}$. For general N-dimensional regions the theorem was proved in an extremely ingenious though quite elementary way by Weinberger $\begin{bmatrix} 99 \end{bmatrix}$.

From Theorem I and Theorem VI it follows that for general R_N

$$(5.19) \qquad \mu_2 < \lambda_1$$

an inequality first observed by Polya $\begin{bmatrix} 70 \end{bmatrix}$. It has been shown in fact (see Payne $\begin{bmatrix} 51 \end{bmatrix}$) that if in two dimensions R_2 is convex

$$(5.20) \qquad \mu_{n+2} < \lambda_n , \qquad n \geq 1 .$$

L. E. Payne

Isoperimetric inequalities for the first non-zero eigenvalue of (E) have been given by Weinstock $[106]$ and Payne $[52]$. Since these inequalities yield upper bounds for q_2 they are not helpful in the applications treated in sections (VI-VIII) and hence will not be discussed here.

Inequalities for eigenvalues of elastically supported membranes and plates, for those of inhomogeneous membranes and for those arising in various other physical problems can be found in the literature (see e. g., $[29\text{-}35,\ 48\text{-}50,\ 53,57,\ 78\text{-}80,\ 87]$).

We mention finally, before passing on to other considerations that Payne, Polya and Weinberger $[55]$ have shown that for N = 2,

$$\text{(5.21)} \qquad \lambda_{n+1} \leqslant 3\lambda_n\ ,$$

$$\text{(5.22)} \qquad \Omega_{n+1} \leqslant 9\Omega_n$$

$$\text{(5.23)} \qquad \Lambda_2 \leqslant 3\Lambda_1\ .$$

The authors have conjectured that the optimal constant in (5.21) is really 2.539, the ratio of λ_2 / λ_1 for a circular membrane.

L. E. Payne

VI.   An Isoperimetric Inequality in Classical Elasticity:

We derive in this section a lower bound for the fundamental vibration frequency of an incompressible isotropic elastic body occupying a region R_3 and fixed on its boundary C_3. This bound yields an important criterion for stability of viscous fluid flow (see e. g., Serrin $\begin{bmatrix} 89 \end{bmatrix}$). It improves the previous criterion of Serrin $\begin{bmatrix} 89 \end{bmatrix}$ and overlaps a recent result of Velte $\begin{bmatrix} 98 \end{bmatrix}$. The theorem which we shall prove in this section is summarized in a recent note of Payne and Weinberger $\begin{bmatrix} 61 \end{bmatrix}$.

We omit the details of the derivation of the governing equations (referring the reader to a text on the Theory of Elasticity) and merely state that the eigenvalue ω (which is proportional to the square of the fundamental frequency) is characterized by the following minimum principle:

$$(6.1) \qquad \omega = \min \frac{\sum_{i,j=1}^{3} \int_{R_3} \varphi_{i,j}\, \varphi_{i,j}\; dv}{\sum_{i=1}^{3} \int_{R_3} \varphi_i\, \varphi_i\; dv}.$$

The minimum is taken among all piecewise continuously differentiable vector fields φ_i which satisfy

$$(6.2) \qquad \begin{aligned} \sum_{i=1}^{3} \varphi_{i,i} &= 0 \text{ in } R_3 \\ \varphi_i &= 0 \text{ on } C_3 \quad , \qquad i = 1,\, 2,\, 3. \end{aligned}$$

It is an immediate consequence of (6.1) and (6.2) that ω is a non-increasing functional of the domain R_3. Thus ω may be bounded below by the corresponding eigenvalue of any domain containing R_3. We compute the exact value for ω if R_3 is a sphere of diameter d. This value will then be a lower bound for the fundamental frequency of any do-

L. E. Payne

main in terms of the diameter d of the smallest circumscribed sphere about R_3. We prove in fact the following theorem:

Theorem VII. <u>The first eigenvalue ω of a vibrating incompressible elastic medium fixed on the boundary is not less than the second eigenvalue in the fixed 3-dimensional membrane problem for any circumscribing sphere.</u>

Now for the sphere R_3 we let U_i be the vector field which minimizes the right hand side of (6.1) subject to conditions (6.2). It is established by the usual arguments of the calculus of variations that U_i exists, is twice continuously differentiable in R_3 and satisfies there the set of Euler equations

$$(6.3) \qquad \Delta U_i + \omega U_i = \partial h / \partial x_i$$

$$(6.4) \qquad U_{i,i} = 0$$

and

$$(6.5) \qquad U_i = 0$$

on the boundary of the sphere. In (6.3) h is an auxiliary function. From (6.3) and (6.4) it follows that

$$(6.6) \qquad \Delta h = 0$$

We now define the function

L. E. Payne

$$(6.7) \qquad V = \sum_{i=1}^{3} x_i U_i$$

Using (6.3) and (6.4) we obtain

$$(6.8) \qquad \Delta V + \omega V = \sum_{i=1}^{3} x_i \partial h / \partial x_i$$

or

$$(6.9) \qquad \Delta (\Delta + \omega) V = 0$$

Moreover, since V vanishes on the boundary of th sphere, and the divergence of U_i vanishes throughout R_3 it follows that

$$(6.10) \qquad V, \quad \frac{\partial V}{\partial n} = 0$$

on the boundary of the sphere. Thus either $V \cong 0$ or ω is an eigenvalue Λ of (B). But by Theorem IV it follows that either $V \cong 0$ or $\omega \geqslant \lambda_2$. (Actually the eigenfunction corresponding to ω cannot be the first eigenfunction of (B) since V must vanish at the origin in view of (6.7)).

Thus either $V \cong 0$ or the theorem is proved.

If $V \equiv 0$ it is clear from (6.8) that $\sum_{i=1}^{3} x_i \frac{\partial h}{\partial x_i} = 0$ in R_3. Since h is a regular harmonic function it must then be a constant. Thus, the components U_i satisfy

$$(6.11) \qquad \Delta U_i + \omega U_i = 0$$

in R_3, and vanish on the boundary. Not all of the U_i may vanish iden-

L. E. Payne

tically. Consequently U_i must then be an eigenvalue of (A). Since the first eigenvalue of (A) for the sphere is simple it is clear that the divergence condition (6.4) cannot be satisfied if $U_i = c_i u_i$. It follows then that $\omega \geq \lambda_2$.

In order to show that the eigenvalue ω for the sphere is equal to λ_2 we note that the vector field

$$U_1 = x_2 \, J_{3/2} \, (\sqrt{\omega} \; r) \Big/ r^{3/2}$$

(6.12)
$$U_2 = -x_1 \, J_{3/2} \, (\sqrt{\omega} \; r) \Big/ r^{3/2}$$

$$U_3 = 0$$

satisfies (6.4), (6.5), and (6.11) provided

(6.13)
$$\omega = 4 \, p^2 / d^2$$

where p is the lowest positive root of the equation

(6.14)
$$\tan p = p.$$

In (6.13) r denotes the distance from the origin and J denotes the Bessel function. The ω of (6.13) is precisely equal to λ_2 for the sphere and Theorem VII is proved. The results of this section are directly applicable in Stokes flow problems for a viscous fluid. The interested reader may find the necessary formulation in any standard text on Fluid Dynamics.

L. E. Payne

VII. Application of Isoperimetric Inequalities in the Study of the Nodal Domains of (A).

In this section we discuss some interesting results obtained by Pleijel $\begin{bmatrix} 64 \end{bmatrix}$, through use of the Faber-Krahn inequality and inequalities (5.19) and (5.20).

Let φ be a twice continuously differentiable function defined on a region R_N. A connected domain T is called a nodal domain of φ if $\varphi \neq 0$ in T and if T is bounded by surfaces $\varphi = 0$ and perhaps also portions of the boundary C_N. For simplicity we consider only the two-dimensional case, and make the further assumption that R_2 be simply connected. A similar result will hold in higher dimensions.

We make use of the Courant nodal line theorem (14) which states that <u>the number of nodal domains of an eigenfunction belonging to the nth eigenvalue of (A) (or (B)) is less than or equal to n.</u> From this theorem and inequality (5.19) it follows that the eigenvalue V_2 corresponding to the μ_2 of (B) cannot have a closed nodal line. For if V_2 had a ring nodal line, then it would be the continuation into R_2 of the first eigenfunction u_1 of the region T bounded by the ring nodal line, i. e., $\mu_2 = \lambda_1^T$ where λ_1^T denotes the first eigenvalue in the fixed membrane problem for T. But, by the well-known monotony principle for the eigenvalues of (A) it follows that $\lambda_1^T \geq \lambda_1$, which would give

$$(7.1) \qquad\qquad \mu_2 \geq \lambda_1$$

in contradiction to (5.19). Thus V_2 cannot have a closed nodal line. By (5.10) it follows in the same way that if R_2 is convex then V_3

140

L. E. Payne

cannot have a closed nodal line. In fact in this case the nodal lines of V_3 will consist of one or two transverses not cutting each other. The inequality (5.10) shows further that for convex R_2, V_n can have at most n-2 "interior" nodal domains.

The question as to whether u_2 can have a closed nodal line has not as yet been answered. If it could be established that u_2 can not have a ring nodal line (and there appears to be reason to believe that this is true) then a number of additional isoperimetric inequalities would follow.

We prove now the following theorem due to Pleijel $\begin{bmatrix} 64 \end{bmatrix}$.

Theorem VIII.   For only a finite number of eigenvalues u_n of (A) will the number of nodal domains of u_n be equal to n.

To prove this theorem we let $T_1, T_2, \ldots\ldots, T_\sigma$ be the nodal domains of the eigenfunction u_n corresponding to λ_n of (A). In each T_i the function $u_n \neq 0$, so that λ_n is the first eigenvalue of a membrane covering T_i . Hence by the Faber-Krahn inequality (3.13) we have

$$(7.2) \qquad A_i / \pi j_o^2 \geq \lambda_n^{-1}$$

where A_i denotes the area of the ith nodal domain. By adding the inequalities we have

$$(7.3) \qquad A / \pi j_o^2 \geq \sigma \lambda_n^{-1}$$

or

L. E. Payne

$$(7.4) \qquad \frac{\lambda_n}{n} \ \frac{A}{\pi j_0^2} \ \geqslant \ \sigma/n \ .$$

Taking the limit as $n \to \infty$ and noting that (Wyl's law $\begin{bmatrix}107, 108\end{bmatrix}$) λ_n/n tends to $4\pi/A$ we obtain

$$(7.5) \qquad \lim_{n \to \infty} \sup \ \frac{\sigma}{n} \ \leqslant \ \frac{\pi}{j_0^2} \ = \ 0.691\ldots,$$

which proves the theorem of Pleijel. In the cases of the square and the circle the maximal subdivision by nodal domains occurs only for $n = 1, 2,$ and 4.

The result of Pleijel has been extended by Peetre $\begin{bmatrix}62\end{bmatrix}$, $\begin{bmatrix}63\end{bmatrix}$, to Riemanian Manifolds. The result may also be extended in the following way (see for instance Bramble and Payne $\begin{bmatrix}12\end{bmatrix}$). Let T_i^+ denote any nodal domains of a function φ for which φ is positive, and let T_i^- denote any nodal domain over which φ is negative. Let A_i^+ denote the area of T_i^+ and A_i^- the area of T_i^- . We establish the following theorem:

Theorem IX: If for any constant λ

$$(7.6) \qquad \begin{aligned} \Delta \varphi + \lambda \varphi &\geqslant 0 \qquad \text{in } R_2 \\ \varphi &\leqslant 0 \qquad \text{on } C_2 \end{aligned}$$

then

    a) T_i^+ is empty, if $\lambda < \lambda_1$

    b) T_i^+ is not empty if $\lambda > \lambda_1$

    and for each nodal domain T_i^+ , $A_i^+ \geqslant \pi j_0^2/\lambda$.

L. E. Payne

To prove this theorem we note that either $\varphi \leq 0$ throughout R or T_i^+ will be non-empty, and each T_i^+ will be bounded by a nodal line. Thus for any nodal domain T_i^+ we have

$$(7.7) \qquad \lambda_1(T_i^+) \int_{T_i^+} \varphi^2 \, dv \leq \int_{T_i^+} |\mathrm{grad}\, \varphi|^2 \, dv =$$

$$= - \int_{T_i^+} \varphi \, [\Delta\varphi + \lambda\varphi] \, dv + \lambda \int_{T_i^+} \varphi^2 \, dv \, .$$

The first integral on the right of (7.7) is non-negative. Thus

$$(7.8) \qquad \lambda_1(T_i^+) \int_{T_i^+} \varphi^2 \, dv \leq \lambda \int_{T_i^+} \varphi^2 \, dv.$$

From the monotony principle for λ_1, it follows that

$$(7.9) \qquad \lambda_1(T_i^+) \geq \lambda_1.$$

Thus, if $\lambda < \lambda_1$ the insertion of (7.8) into (7.7) leads to a contradition unless T_i^+ is empty.

This proves the first part of the theorem.

If $\lambda > \lambda_1$ and T_i^+ is not empty we obtain by an application of the Faber-Krahn inequality to (7.7)

$$(7.10) \qquad \frac{\pi \, j_0^2}{A_i^+} \int_{T_i^+} \varphi^2 \, dv \leq \lambda \int_{T_i^+} \varphi^2 \, dv.$$

This proves assertion b) if it can be shown that for $\lambda > \lambda_1$, T_i^+ is non-empty. To prove this we make use of the fact that the first eigenfunction u_1 of (A) is positive throughout R_2. We assume that φ is nowhere

L. E. Payne

positive and show that this leads to a contradiction. From Green's identity we have

$$(7.11) \qquad \int_{R_2} u_1 \left[\Delta \varphi + \lambda \varphi \right] dv = - \oint_{C_2} \varphi \frac{\partial u_1}{\partial n} ds + (\lambda - \lambda_1) \int_{R_2} \varphi u_1 dv .$$

The term on the left is non-negative while if φ is nowhere positive in R_N the terms on the right are non-positive for $\lambda > \lambda_1$. Hence, if we exclude the trivial solution $\varphi \equiv 0$, it follows that for $\lambda > \lambda_1$, must be positive at some point in R , and the theorem is proved. Similar results have been obtained by Hartman and Wintner $\left[27 \right]$, Protter $\left[80 \right]$ and McNabb $\left[47 \right]$.

Let G denote the number of components T_i^+ . Then

$$(7.12) \qquad A \geq \sum_{i=1}^{G} A_i^+ \cdot \frac{\pi j_0^2 G}{\lambda}$$

or

$$(7.13) \qquad G \leq \lambda A / \pi j_0^2 .$$

This gives an upper bound on the number of nodal domains T_i^+ .

L. E. Payne

VIII.   A Priori Bounds:

In this section we are concerned with boundary value problems for uniformly elliptic operators. We frequently wish to obtain upper and lower bounds for the error in the approximation of the solution (or its dervatives) at points inside some region R_N, if the data of the boundary value problem is approximated in the mean square sense. Often we may be interested only in an L_2 bound for the error (or its derivatives) over the region. Methods for obtaining pointwise bounds and bounds for energy integrals have been known for some time (see Diaz and Weinstein [19] , Prager and Synge [77] , Diaz and Grenberg [18] , Greenberg [25] , Diaz [16, 17] , Maple [46] , Synge [92, 93] , and others). In these various methods the approximating functions must belong to special classes of functions, and the upper and lower bounds require the use of two different classes of functions. For example, lower bounds may require the use of functions satisfying the differential equation while upper bounds involve approximating functions satisfying the boundary conditions.

Considerable labor might be saved if we were able to obtain both upper and lower bounds from a single set of approximating functions, especially if we were able to use as approximating functions any sufficiently smooth but otherwise arbitrary functions. With this in mind, we are lead to seek appropriate a priori bounds. For example, suppose we wished to consider the following boundary value problem for an elliptic operator $L(u)$ of order $2m$ defined on a region R_N with boundary C_N. a) $L(u)$ prescribed in R_N, b) a set of boundary conditions $B_i(u)$ $(i = 1, \ldots, m-1)$ prescribed on C_N. Let us suppose that we are able to obtain an a priori bound of the following type for an arbitrary sufficiently smooth function W.

L. E. Payne

(The functions u and w employed in this section are not to be confused with the eigenfunctions of (A) and (B)).

$$(8.1) \qquad \int_{R_N} w^2 dv \; \leqslant \; \sum_{i=1}^{m-1} k_i \oint_{C_N} B_i^2(w) ds + k_m \int_{R_N} \left[L(w) \right]^2 dv,$$

with explicitly determined constants K_j . Then by taking $W = u - \varphi$ where φ is any sufficiently smooth approximating function we would have an L_2 bound for the error in approximating u in terms of mean square integrals of the error in approximation of the data. The Rayleigh-Ritz technique could then be used in determining an optimal choice for φ (among linear combinations of a given set of functions).

By the triangle inequality it follows that

$$(8.2) \quad \left[\int_{R_N} \varphi^2 dv \right]^{1/2} - \left[\int_{R_N} w^2 dv \right]^{1/2} \leqslant \left[\int_{R_N} u^2 dv \right]^{1/2} \leqslant \left[\int_{R_N} \varphi^2 dv \right]^{1/2} +$$
$$+ \left[\int_{R_N} w^2 dv \right]^{1/2} .$$

We would thus have upper and lower bounds for the L_2 integral of the solution u . This demonstrates that once explicit constants K_j have been obtained, the L_2-bounds follow in a straight forward manner.

If we had in addition an inequality

$$(8.3) \qquad \left| w(P) \right|^2 \leqslant C_0(P) \int_{R_N} w^2 dv + F\big(L(w) \big)$$

where F denotes some known functional of L(W) and $C_0(P)$ is an explicitly determined constant, then (8.3) and (8.1) together would give

L. E. Payne

pointwise bounds for u.

The problem of finding a priori bounds is thus reduced to the problem of obtaining the explicit constants. The existence of such constants is assured for quite a wide class of boundary value problems, but explicit constants are often difficult to compute. The optimal constants are usually just eigenvalues of associated eigenvalue problems. For istance, suppose all of the $B_i(w)$ vanish on C. Then if we define for sufficiently smooth ψ

$$(8.4) \qquad \alpha_1 = \min_{B_i(\psi) = 0 \text{ on } C_N} \frac{\int_{R_N} [L(\psi)]^2 dv}{\int_{R_N} \psi^2 dv}$$

Then (assuming $\alpha_1 \neq 0$) the optimal choice for K_m is α_1^{-1}. The constant α_1 will usually not be known. However, we may use any lower bound for α_1 as our K_m. In particular, if we have some isoperimetric inequality, which gives a lower bound for α_1, we may use this value for our K_m.

Consider the special case in which L is the Laplace operator and $B_1(\psi) = \psi$. It is well known that

$$(8.5) \qquad \lambda_1^2 = \min_{\psi = 0 \text{ on } C_N} \frac{\int_{R_N} (\Delta \psi)^2 dv}{\int_{R_N} \psi^2 dv} \quad .$$

This leads then to the inequality for functions W which vanish on C

$$(8.6) \qquad \int_{R_N} w^2 dv \leq \frac{1}{\lambda_1^2} \int_{R_N} (\Delta w)^2 dv \leq \frac{1}{\lambda_1^2} \int_{R_N} (\Delta w)^2 dv$$

where $\bar{\lambda}_1$ is any lower bound for λ_1.

L. E. Payne

Let us now take for $L(u)$

$$(8.7) \qquad L(u) = (a^{ij}(x)u_{,i})_{,j} \ , \qquad x = (x_1, x_2, \ldots, x_N) \ ,$$

where the summation convention is understood, and the comma denotes partial differentiation. The components of the symmetric matrix a^{ij} are assumed to be piecwise continuously differentiable in R_N. It is assumed further that positive constants a_0 and a_1 exist such that for all real numbers $(\xi_1, \xi_2, \ldots \xi_N)$ and all x in R_N:

$$(8.8) \qquad a_0 \sum_{i=1}^{N} \xi_i^2 \ \le \ a^{ij}\xi_i\xi_j \ \le \ a_1 \sum_{i=1}^{N} \xi_i^2 \ .$$

The idea of using a priori inequalities for obtaining bounds was used by Fichera [21, 22] who derived on a priori inequality of the following type (see also Bramble ane Payne [10]):

$$(8.9) \qquad \left[\int_{R_N} w^2 \, dv\right]^{1/2} \le \left[k \oint_{C_N} w^2 \, ds\right]^{1/2} + \left[\frac{1}{a_0}\lambda_1^2 \int_{R_N} \left[L(w)\right]^2 dv\right]^{1/2} .$$

The optimal constant K is the reciprocal of the eigenvalue τ defined by the following minimum principle (see e.g. Fichere [22] , Bramble and Payne [10])

$$(8.10) \qquad \tau = \min_{\psi = 0 \text{ on } C_N} \frac{\int_{R_N} \left[L(\psi)\right]^2 dv}{\int_{C_N} \left(\frac{\partial\psi}{\partial v}\right)^2 ds}$$

L. E. Payne

where $\partial/\partial v$ denotes the conormal derivate, i. e.,

(8.11) $$\frac{\partial}{\partial v} = a^{ij} \frac{\partial u}{\partial x_i} n_j .$$

Under sufficient smoothness assumptions the minimizing function V satisfies

$$L L (v) = 0 \qquad \text{in } R_N$$

(8.12) $$\left. \begin{array}{l} v = 0 \\[2mm] L (v) - \tau \dfrac{\partial v}{\partial v} = 0 \end{array} \right\} \text{on } C_N$$

τ is a Stekloff type eigenvalue. Although an isoperimetric inequality is known for λ_1 , no such isoperimetric inequality is known for τ . Results of Payne and Weinberger $\begin{bmatrix} 56 \end{bmatrix}$ do, however, give a crude lower bound for τ which is sufficient for our purposes. Methods which in some cases give very good bounds for τ were also introduced by Fichera $\begin{bmatrix} 21 \end{bmatrix}$.

For points P on the interior of R_N an explicit constant $C_o(P)$ and a function F can be obtained (see e. g. $\begin{bmatrix} 10 \end{bmatrix}$) without assuming existence of the Green's function. The insertion of the explicit constants in (8.3) and (8.9) then yields the appropriate bounds. We should point out that similar results are contained in papers of Fichera $\begin{bmatrix} 21\text{-}23 \end{bmatrix}$, and Payne and Weinberger $\begin{bmatrix} 56 \end{bmatrix}$.

We consider now the Neumann problem for L(u) i. e., a) L(u) prescribed in R_N and b) $\frac{\partial u}{\partial v}$ prescribed on the boundary C_N of R_N . Let us suppose first that the surface C_N is convex. We wish to

L. E. Payne

obtain pointwise bounds for u and its derivatives in R_N. Since the solution is determined only up to an additive constant, we fix the latter by the normalization

$$(8.13) \qquad \int_{R_N} (u - \varphi) \, dv = 0 ,$$

where φ is a prescribed approximating function.

By inequality (4.13) we have for any sufficiently smooth function W satisfying $\int_{R_N} w \, dv = 0$ in R_N

$$(8.14) \qquad \int_{R_N} w^2 \, dv \le \frac{D^2}{\pi^2} \int_{R_N} |\operatorname{grad} w|^2 \, dv \le \frac{D^2}{\pi^2 a_c} \int_{R_N} a^{ij} w_{,i} w_{,j} \, dv .$$

We seek now a bound for $\oint_{C_N} w^2 \, ds$.

Since R_N is convex it follows that if the origin is taken at some point in R_N then $t \equiv x_i n_i > 0$ at every point on C_N. Consider then

$$(8.15) \qquad \oint_{C_N} x_i n_i w^2 \, ds = N \int_{R_N} w^2 \, dv + 2 \int_{R_N} x_i w w_{,i} \, dv .$$

By the arithmetic geometric mean inequality

$$(8.16) \qquad \oint_{C_N} t w^2 \, ds \le (N + \alpha) \int_{R_N} w^2 \, dv + D^2 \alpha^{-1} \int_{R_N} |\operatorname{grad} w|^2 \, dv$$
$$\le \left[\frac{N + \alpha}{\pi^2} + \frac{1}{\alpha} \right] \frac{D^2}{a_0} \int_{R_N} a^{ij} w_{,i} w_{,j} \, dv, \qquad \alpha > 0$$

L. E. Payne

The optimal choice of α ($\alpha = \pi$) gives the result

$$(8.17) \qquad \oint_{C_N} t\, w^2 ds \;\leqslant\; \frac{(N + 2\pi)}{\pi\, a_0}\, D^2 \int_{R_N} a^{ij}\, w_{,i}\, w_{,j}\, dv \;.$$

It can be shown without difficulty that a lower bound for the first non-zero eigenvalue q_2 of E follows immediately from (8.17) i.e.

$$(8.18) \qquad q_2 \;>\; \frac{\pi\, t\, \min .}{(N + 2\pi)\, D^2} \;.$$

This inequality is however, not isoperimetric.

Let us now obtain a bound for $\displaystyle\int_{R_N} a^{ij}\, w_{,i}\, w_{,j}\, dv$ in terms of the Neumann data of W. From Green's identity we have

$$(8.19) \qquad \int_{R_N} a^{ij}\, w_{,i}\, w_{,j}\, dv = \oint_{C_N} w\, \frac{\partial w}{\partial \nu}\, ds - \int_{R_N} w\, L(w)\, dv \;.$$

Then by Schwarz's inequality we obtain

$$(8.20) \qquad \int_{R_N} a^{ij} w_{,i} w_{,j}\, dv \;\leqslant\; \left[\oint_{C_N} t\, w^2 ds \oint_{C_N} t^{-1}\left(\frac{\partial w}{\partial \nu}\right)^2 ds\right]^{1/2} + \left[\int_{R_N} w^2 dv \int_{R_N} L(w)^2 dv\right]^{1/2}$$

and from (8.14) and (8.17),

$$(8.21) \qquad \left[\int_{R_N} a^{ij} w_{,i} w_{,j}\, dv\right]^{1/2} \;\leqslant\; \left[\frac{(N + 2\pi)D^2}{\pi^2 a_0} \oint_{C_N} t^{-1}\left(\frac{\partial w}{\partial \nu}\right)^2 ds\right]^{1/2} +$$

$$+ \left[\frac{D}{\pi^2 a_0} \int_{R_N} \left[L(w)\right]^2 dv\right]^{1/2} \;.$$

L. E. Payne

Inequality (8.21) together with (8.14) and (8. 3) thus yields an upper bound for $W(P)$. By setting $W = u - \varphi$ we obtain the desired pointwise bounds for u. Using similar techniques it is also possible to obtain bounds for derivatives of u.

We have obtained a simple bounds in the Neumann problem for a convex domain. We wish, however, to treat the Neumann problem for more general regions. It is clear that convexity was used only in establishing (8.14) and (8.17). Thus the critical step in the derivation of bounds in the Neumann problem for a general region is the establishment of the corresponding inequalities (8.14) and (8.17). We show now how a lower bound $\oint$ for q_2 of (E) leads to the desired inequalities.

Since $\oint$ is a lower bound for q_2 then for $\tilde{w}$ normalized so that $\int_{C_N} \tilde{w}\, ds = 0$ (Note that we are now using a different normalization than that used previously, i.e., $\tilde{w} = w + \text{constant}$).

$$(8.22) \qquad \oint_{C_N} \tilde{w}^2\, ds \leq (\oint a_\rho)^{-1} \int_{R_N} a^{ij}\tilde{w},_i\, \tilde{w},_j\, dv \ .$$

In (8.15) with w replaced by $\tilde{w}$ we use the arithmetic-geometric mean inequality to solve for $\int_{R_N} \tilde{w}^2\, dv$ in terms of $\oint_{C_N} \tilde{w}^2\, ds$ and $\int_{R_N} a^{ij}\tilde{w},_i\, \tilde{w},_j\, dv$, i.e.

$$(8.23) \qquad \int_{R_N} w^2\, dv \leq \frac{2D}{N} \oint_{C_N} \tilde{w}^2\, ds + \frac{1}{a_o}\left(\frac{2D}{N}\right)^2 \int_{R_N} a^{ij}\tilde{w},_i\, \tilde{w},_j\, dv$$

$$\leq \frac{2D}{Na_o}\left[\oint^{-1} + \frac{2D}{N}\right] \int_{R_N} a^{ij} w,_i\, w,_j\, dv.$$

For general regions Bramble and Payne [11] have obtained a lower bound

L. E. Payne

$\int$ for q_2. This gives not only the desired bounds in the Neumann problem, but also a lower bound for the μ_2 of (B).

We have illustrated by some simple esamples how the optimal constants in our a priori bounds are related to the eigenvalues of various problems. Many more esamples could be given but let me conclude by considering a some-what different type of problem.

We seek bounds for the solution w of the Dirichlet problem for the operator $\Delta w + V w$ where V is a constan lying between λ_n and λ_{n+1} of (A). Bramble and Payne $[12]$ have computed a priori bounds of the following type for an arbritary sufficiently smooth function w .

$$(8.24) \qquad \int_{R_N} w^2 \, dv \leq K_1 \sum_{i=1}^{n+1} \frac{1}{(\lambda_i - V)^2} \left\{ \oint_{C_N} w^2 \, ds + \beta \int_{R_N} [\Delta w + V w]^2 \, dv \right\}$$

where the constants K_1 and β are explicit.

If we knew the eigenvalues λ_i we would then have an a priori bound for $\int_{R_N} w^2 \, dv$ in terms of the Dirichlet data. If the λ_i are not known it suffices to have a lower bound for λ_{n+1} which is still larger than V and an upper bound for λ_n which is still smaller than V . The upper bounds are usually obtained from an application of the Rayleigh-Ritz techique to the Rayleigh quotient. Upper bounds are somewhat more difficult. However, various methods for obtaining lower bounds are known (see e.g., Weinstein $[103, 104]$, Aronszajn $[1]$, Temple $[97]$, Kato $[37]$, Bazley $[4, 5]$, Weinberger $[102]$, Bazley and Fox $[6]$, and others $[2, 13, 26, 42]$.) These and other methods are discussed in the papers of Weinstein and De Vito which appear in this vo-

L. E. Payne

lume and will not be considered here.

It is again possible to derive, for points interior to R_N, the inequality,

$$(8.25) \qquad \left| W(P) \right|^2 \leq C_1(P) \int_{R_N} w^2 dv + F_1 (\Delta w + \nabla w)$$

with explicit $C_1(P)$ and F_1. This leads then to pointwise bounds for u.

We have considered only a simple example of a forced vibration-type problem. Much more general results have been obtained. (see $\begin{bmatrix} 12 \end{bmatrix}$).

L. E. Payne

IX. <u>Concluding Remarks</u> :

In this paper we have presented a few of the most interesing and most useful isoperimetric inequalities for eigenvalues. The many important isoperimetric inequalities for energy integrals (torsional rigidity, electrostatic capacity, virtual mass, polarization, etc.) have not been considered. Eigenvalue inequalities have then been used to investigate various properties of eigenfunctions and solutions to boundary value problems. They have been employed finally in the determination of a priori bounds for solutions to various boundary value problems.

The bibliography which follows is not complete, but is intended only to be representative. Additional references may be obtained from the bibliographies of the books and papers cited there.

L. E. Payne

Bibliography

1. Aronszajn, N., <u>Approximation methods for eigenvalues of completely continuous symmetric operators</u>, Symp. Spectral Theory and Diff. Probs, Stillwater, Oklahoma (1951) pp. 179-202.

2. Aronszajn, N., and Weinstein, A., <u>On a nified theory of eigenvalues of plates and membranes</u>, Amer. J. Math., vol. 64 (1942) pp. 623-645.

3. Banks, D., <u>Bounds for the eigenvalues of some vibrating systems</u>, Pac. J. Math., vol. 10 (1960) pp. 439-474, see also Pac. J. Math., vol. 11 (1961) pp. 1183-1203.

4. Bazley, N., <u>Lower bounds for eigenvalues with applications to the helium atom</u>, Proc. Nat'l Acad. Sci, vol. 45 (1959) pp. 144-149.

5. Bazley, N., <u>Lower bounds for eigenvalues</u>, J. Math. Mech., vol. 10 (1961) pp. 289-308.

6. Bazley, N., and Fox, D., <u>Truncations in the method of intermediate problems for lower bounds for eigenvalues</u>, J. Res. Natl. Bureau Standards, vol. 65 B, (1961) pp. 105-111.

7. Beesack, P. R., <u>A note on an integral inequality</u>, Proc. Amer. Math. Soc., vol. 8 (1957) pp. 875-879.

8. Beesack, P. R., <u>Isoperimetric inequalities for the nonhomogeneous clamped rod and plate</u>. J. Math. and Mech., vol. 8 (1959) pp. 471-482.

9. Beesack, P. R. and Schwarz, B., <u>On the zeros of solutions of second-order linear differential equations</u>, Can. J. Math., vol. 8 (1956) pp. 504-515.

10. Bramble, J. H., and Payne, L. E., <u>Bounds for solutions of second order partial differential equations</u>. Contrib. to Diff. Eqtns.

L. E. Payne

(to appear).

11. Bramble, J. H., and Payne, L. E., Bounds in the Neumann problem for second order uniformly elliptic operators. Pac J. Math. (in print.)

12. Bramble, J. H., and Payne, L. E., Upper and lower bounds in forced vibration and allied problems. (to appear).

13. Collatz, L., Eigenwertprobleme und ihre numerische Behandlung, Chelsea Press, New York (1948).

14. Courant, R., Ein allgmeiner Satz zur Theorie der Eigenfunktionen selbstadjungierter Differential aus drücke, Nach Akad. Wiss. Göttingen (1923) pp. 81-84.

15. Courant, R., and Hilbert, D., Methoden der Mathematischen Physik, vol. 1, Springer, Berlin (1931). English Edit. Methods of Mathematical Phisics, vol. 1, Interscience, New York (1953).

16. Diaz, J. B., Upper and lower bounds for quadratic functionals, Proc. Symp. Spectral Theory and Diff. Probs. Oklahoma A. & M. (1950) pp. 279-289, see also Collectaneae Math., vol. 4 (1951) pp. 3-50.

17. Diaz, J. B., Upper and lower bounds for quadratic integrals, and at a point, for solutions of linear boundary value problems, Proc. Symp. Bdry. Val. Probs. Diff. Eqtns., U. S. Army Research Center, Univ. Wisconsin, April (1959) pp. 47-83.

18. Diaz, J. B., and Greenberg, H. J., Upper and lower bounds for the solution of the first biharmonic boundary value problem, J. Math. Phys., vol. 27 (1948) pp. 193-201.

19. Diaz, J. B., and Weinstein, A., Scharz's inequality and the me-

L. E. Payne

thods of Rayleigh-Ritz and Trefftz, J. Math. Phys. vol. 26, (1947) pp. 133-136.

20. Faber, G. , Beweis, dass unter aller homogenen Membranen von gleicher Fläche und gleicher Spannung die Kreisförmige den tiefsten Grundton gibt, Sitz. bayer. Akad. Wiss. (1923) pp. 169-172.

21. Fichera, G. , Formule di maggiorazione connesse ad una classe di transformazioni lineari, Annali Mat. Pura Appl. vol. 36 (1954) pp. 273-296.

22. Fichera, G. , Methods of functional linear analysis in mathematical physics, Proc. Int. Cong. Math. , Amsterdam, vol. 3 (1954) pp. 216-228.

23. Fichera, G. , Alcuni recenti sviluppi della teoria dei problemi al contorno per le equazioni alle derivate parziali lineari, Conv. Inter. Equaz. Lin. Alle Deriv. Parz. (1954) Trieste.

24. Forsythe, G. Asymptotic lower bounds for the frequencies of certain polygonal membranes, Pac. J. Math. , vol. 4 (1954) pp. 467-480.

25. Greenberg, H. J. , The determination of upper and lower bounds for the solution of the Dirichlet problem, J. Math. Phys. , vol. 27 (1948) pp. 161-182.

26. Gould, S. H. , Variational Methods for Eigenvalue problems, Univ. Toronto Press (1957).

27. Hartman, P. , and Wintner, A. , On a comparison theorem for self-adjoint partial differential equations of elliptic type, Proc. Amer. Math. Soc. , vol. 6 (1955) pp. 862-865.

L. E. Payne

28.   Hersch, J., Equations différentielles et fonctions de cellules, C. R. Acad. Sci. Paris, vol. 240 (1955) pp. 1602-1604.

29.   Hersch, J., Un principe de maximum pour la fréquence fondamentale d'une membrane, C. R. Acad. Sci. Paris, vol. 249 (1959) pp. 1074-1076.

30.   Hersch, J., Une méthode pour l'evaluation par défaut de la première valeur de la vibration ou du flambage des plaques encastrées, C. R. Acad. Sci. Paris, vol. 250 (1959) pp. 3943-3945.

31.   Hersch, J., Une interpretation du principe de Thomson et son analogue pour la fréquence fondamentale d'une membrane, C. R. Acad. Sci. Paris, vol. 248 (1959) pp. 2060-2062.

32.   Hersch, J., Sur la fréquence fondamentale d'une membrane vibrante: évaluations par défaut et principe de maximum, ZAMP, vol. 11 (1960) pp. 387-413.

33.   Hersch, J., Physical interpretation and strengthening of M. H. Protter's method for vibrating nonhomogeneous membranes; its analogue for Schorödinger's equation, Pac. J. Math., vol. 11 (1961) pp. 971-980.

34.   Hersch, J., and Payne, L. E., L'éffet d'une contrainte rectiligne sur la frequence fondamentale d'une membrane vibrante, C. R. Acad. Sci. Paris, vol. 249 (1959) pp. 1855-1857.

35.   Hooker, W., and Protter, M. H., Bounds for the first eigenvalue of a rhombic membrane, J. M. Phys., vol. 39 (1960) pp. 18-34.

36.   Hubbard, B., Bounds for eigenvalues of the free and fixed membrane by finite difference methods, Pac. J. Math., vol. 11 (1961) pp. 559-590.

L. E. Payne

37. Kato, T., On the upper and lower bounds for eigenvalues, J. Phys. Soc. Japan, vol. 4 (1949) pp. 415-438.

38. Keller, J. B., The shape of the strongest column, Arch. Rat. Mech. Anal., vol. 5 (1960) pp. 275-285; see also Tadjbakhsh, I., and Keller, J. B., Strongest columns and isoperimetric inequalities for eigenvalues, J. Appl. Mech. vol. 29 (1962) pp. 159-164.

39. Keller, J. B., Lower bounds and isoperimetric inequalities for eigenvalues in the Schrödinger equation, J. Math. Phys., vol. 2 (1961) pp. 262-266.

40. Kornhauser, E. T., and Stakgold, I., A variational theorem for $\nabla^2 u + \lambda u = 0$ and its applications, J. Math. Phys., vol. 31 (1952) pp. 45-54.

41. Krahn, E., Über eine von Rayleigh formulierte Minimaleigenschaft des Kreises, Math. Ann., vol. 94 (1924) pp. 97-100; see also Über Minimaleigenschaft der Kugel in drei und mehr Dimensionen Acta Comm. Univ. Dorp., vol. A 9 (1926) pp. 1-44 .

42. Krein, M. G., On certain problems on the maximum and minimum of characteristic values and on the Lyapunov zones of stability, Amer. Math. Soc. Trans., series 2, vol. 1 (1955) pp. 163-187.

43. Krylov, N., Les méthodes de solution approchée des problèmes de la physique mathématique, Mem. Sci. Math., No. 49 (1931).

44. Makai, E., On the principal frequency of a convex membrane and related problems, Czech. Math. J., vol. 9 (1959) pp. 66-70.

45. Makai, E., Bounds for the principal frequency of a membrane and the torsional rigidity of a beam, Acta Szeged, vol. 20 (1959) pp. 33-35.

46. Maple, C. G., The Dirichlet problem: Bounds at a point for the

L. E. Payne

solution and its derivatives, Quart. Appl. Math., vol. 8 (1950) pp. 213-228.

47. McNabb, A., Strong comparison theorems for elliptic equations of second order, J. Math. Mech., vol. 10 (1961) pp. 431-440.

48. Nehari, Z., On the principal frequency of a membrane, Pac. J. Math., vol. 8 (1958) pp. 285-293.

49. Nehari, Z., Oscillation criteria for second order linear differential equations, Trans. Amer. Math. Soc., vol. 85 (1957) pp. 428-445.

50. Nehari, Z., Some eigenvalue estimates, J. Analy. Math., vol. 7 (1959) pp. 79-88.

51. Payne, L. E., Inequalities for eigenvalues of membranes and plates, J. Rat. Mech. Anal, vol. 4 (1955) pp. 517-528.

52. Payne, L. E., New isoperimetric inequalities for eigenvalues and other physical quantities, Comm. Pure Appl. Math., vol. 9 (1956) pp. 531-542.

53. Payne, L. E., Inequalities for eigenvalues of supported and free plates, Quart. Appl. Math., vol. 16 (1958) pp. 111-120.

54. Payne, L. E., A note on inequalities for plate eigenvalues, J. Math. Phys. vol. 39 (1960) pp. 155-159.

55. Payne, L. E., Polya, G., and Weinberger, H. F., On the ratio of consecutive eigenvalues, J. Math. Phys. vol. 35 (1956) pp. 289-298.

56. Payne, L. E., and Weinberger, H. F., New bounds for solutions of second order partial differential equations, Pac. J. Math., vol. 8 (1958) pp. 551-573.

L. E. Payne

57. Payne, L. E., and Weinberger, H. F., Lower bounds for vibration frequencies of elastically supported membranes and plates, J. Soc. Ind. Appl. Math., vol. 5 (1957) pp. 171-182.

58. Payne, L. E., and Weinberger, H. F., A Faber-Krahn inequality for wedge-like domains, J. Math. Phys., vol. 39 (1960) pp. 182-188.

59. Payne, L. E., and Weinberger, H. F., An optimal Poincaré inequality for convex domains, Arch. Rat. Mech. anal, vol. 5 (1960) pp. 286-292.

60. Payne, L. E., and Weinberger, H. F., Some isoperimetric inequalities for membrane frequencies and torsional rigidity, J. Math. Anal. Appl., vol. 2 (1961) pp. 210-216.

61. Payne, L. E., and Weinberger, H. F., A stability bound for viscous flows, Symp. on Non-linear Problems, Math. Res. Cent. U. S. Army, Univ. of Wisc. (1962).

62. Peetre, J., A generalization of Courant's nodal line theorem, Math. Scand., vol. 5 (1957) pp. 15-20.

63. Peetre, J., Estimates of the number of nodal domains, Proc. 13 Cong. Math. Scand. (1957) pp. 198-201.

64. Pleijel, Å., Remarks on Courant's nodal line theorem, Comm. Pure Appl. Math., vol. 9 (1956) pp. 543-550.

65. Poincaré, H., Sur les équations aux derivées partielles de la physique mathématique, Amer. J. Math., vol. 12 (1890) pp. 259-261.

66. Poincaré H., Figures d' équilibre d'une masse fluide, Paris (1903).

67. Polya, G., Sur la fréquence fondamentale des membranes vibrantes et la résistance élasique des tiges à la torsion, C. R. Acad. Sci. Paris, vol. 225 (1947) pp. 346-348.

L. E. Payne

68.   Polya, G., A minimum problem about the motion of a solid through a fluid, Proc. Nat'l. Acad. Sci. U. S. A., vol. 33 (1947) pp. 218-221.

69.   Polya, G., Torsional rigidity, principal frequency, electrostatic capacity and symmetrization, Quart. Appl. Math., vol. 6 (1948) pp. 267-277.

70.   Polya, G., Remarks on a foregoing paper, J. Math. Phys., vol. 31 (1952) pp. 55-57.

71.   Polya, G., Sur une interprétation de la méthode des différences finies qui peut fournir des bornes superieures ou inférieures, C. R. Acad. Sci. Paris, vol. 235 (1952) pp. 995-997.

72.   Poyla, G., More isoperimetric inequalities proved and conjectural, Comm. Math. Helvetia., vol. 29 (1955) pp. 112-119.

73.   Polya, G., Sur les fréquences propres des membranes vibrantes, C. R. Acad. Sci. Paris, vol. 242 (1956) pp. 708-709; see also Sur quelques membranes vibrantes de forme particulière, ibid, vol. 243 (1956) pp. 469-471.

74.   Polya, G., On the eigenvalues of vibrating membranes, Proc. London. Math. Soc., vol. 11 (1961) pp. 419-433.

75.   Polya, G., Two more inequalities between physical and geometrical quantities, J. Indian Math. Soc., vol. 24 (1960) pp. 413-419.

76.   Polya, G., and Szegö, G., Isoperimetric inequalities in mathematical physics, Annals of Math. Studies No. 27, Princeton U. Press (1951).

77.   Prager, W., and Synge, J. L., Approximations in elasticity based on the concept of function space, Quart. Appl. Math., vol. 5 (1947) pp. 241-269.

L. E. Payne

78. Protter, M. H., Lower bounds for the first eigenvalue of elliptic
     equations, Annals of Math., vol. 71 (1960) pp. 423-444.

79. Protter, M. H., Vibration of a non-homogeneous membrane, Pac.
     J. Math., vol. 9 (1959) pp. 1249-1255.

80. Protter, M. H., A comparison theorem for elliptic equations, Proc.
     Amer. Math. Soc., vol. 10 (1959) pp. 249-299.

81. Lord Rayleigh, The theory of sound, 2nd. ed., London 1884/96.

82. Saint Venant, B. de, Mémoire sur la torsion des prismes, Mem.
     div. Sav. Acad. Sci. vol. 14 (1856) pp. 233-560.

83. Schiffer, M., Sur la polarization et la masse virtuelle, C. R. Acad.
     Sci. Paris, vol. 244 (1957) pp. 3118-3121.

84. Schiffer, M., and Szegö, G., Virtual mass and polarization, Trans.
     Amer. Math. Soc., vol. 67 (1949) pp. 130-205.

85. Schumann, W., On isoperimetric inequalities in plasticity, Quart.
     Appl. Math., vol. 16 (1958) pp. 309-314.

86. Schwarz, B., Bounds for the sums of reciprocals of eigenvalues,
     Bull. Res. Courc. Israel, vol. 8F (1959) pp. 91-102.

87. Schwarz, B., Bounds for the principal frequency of the nonhomo-
     geneous membrane and for the generalized Dirichlet integral,
     Pac. J. Math., vol. 7 (1957) pp. 1653-1676.

88. Schwarz, B., On the extrema of the frequencies of nonhomogeneous
     strings with equimeasurable density, J. Math. Mech., vol. 10
     (1961) pp. 401-422.

89. Serrin, J., On the stability of viscous fluid motions, Arch. Rat.
     Mech. Anal. vol. 3 (1959) pp. 1-13.

90. Steiner, J., Einfache Beweise der isoperimetrischen Hauptsätze,
     Werke II, Berlin, (1882) pp. 75-91.

L. E. Payne

91.   Stekloff, M. W., Sur les problèmes fondamentaux de la physique
        mathématique, Ann. Sci. E'cole Norm. Sup., vol. 19 (1902)
        pp. 455-490.

92.   Synge, J. L., Pointwise bounds for the solutions of certain boun-
        dary value problems, Proc. Roy. Soc. (A), vol. 208 (1951)
        pp. 170-175.

93.   Synge, J. L., The hypercircle in mathematical physics, Cambridge
        U. Press (1957).

94.   Szegö, G., Über einige neue Extremalaufgaben der Potentialtheorie,
        Math. Ziet., vol. 31 (1930) pp. 583-593.

95.   Szegö, G., On the capacity of a condenser, Bull. Amer. Math.
        Soc., vol. 51 (1945) pp. 325-350.

96.   Szegö, G., Inequalities for certain eigenvalues of a membrane of
        given area, J. Rat. Mech. Anal., vol. 3 (1954) pp. 343-356.

97.   Temple, G., and Bickley, W. G., Rayleigh's principle and its ap-
        plications to engineering, Oxford Univ. Press (1933).

98.   Velte, W., Über ein Stabilitätskriterium der Hydrodynamik, Arch.
        Rat. Mech. Anal, vol. 9 (1962) pp. 9-20.

99.   Weinberger, H. F., An isoperimetric inequality for the N-dimen-
        sional free membrane problem, J. Rat. Mech. Anal., vol. 5
        (1956) pp. 533-636.

100.   Weinberger, H. F., Upper and lower bounds for eigenvalues by fi-
        nite difference methods, Comm. Pure Appl. Math., vol. 9 (1956)
        pp. 613-623.

101.   Weinberger, H. F., Lower bounds for higher eigenvalues by finite
        difference methods, Pac. J. Math. vol. 8 (1958) pp. 339-368.

L. E. Payne

102. Weinberger, H. F. , The theory of lower bounds for eigenvalues,
      U. of Md. Tech. Note BN 183 (1959).

103. Weinstein, A. ; Etude des spectres des équations aux dérivées
      partielles de la théorie des plaques élastiques, Mémorial des
      Scien. Math. , vol. 88, Paris (1937).

104. Weinstein, A. , Variational methods for the approximation and
      exact computation of eigenvalues, NBS Applied Math. Series 29
      (1953) pp. 83-89.

105. Weinstein, A. , Generalized axially symmetric potential theory,
      Bull. Amer. Math. Soc. , vol. 59 (1953) pp. 20-38.

106. Weinstock, R. , Inequalities for a classical eigenvalue problem,
      J. Rat. Mech. Anal. , vol. 3 (1954) pp. 745-753.

107. Weyl, H. , Das asymptotische Verteilungsgesetz der Eigenwerte
      linearer partieller Differentialgleichungen (mit einer Anwen-
      dung auf die Theorie der Hohlraumstrahlung), Math. Ann. , vol.
      71 (1912) pp. 441-479.

108. Weyl, H. , Über die Abhangigkeit der Eigenschwingungen einer
      Membram von deren Begrenzung, J. reine Ang. Math. , vol.
      141 (1912) pp. 1-11.

CENTRO INTERNAZIONALE MATEMATICO ESTIVO

(C. I. M. E.)

LUCIANO DE VITO

1. CALCOLO DEGLI AUTOVALORI E DELLE AUTOSOLUZIONI
   PER OPERATORI NON AUTOAGGIUNTI

2. SUL CALCOLO PER DIFETTO E PER ECCESSO DEGLI
   AUTOVALORI DELLE TRASFORMAZIONI HERMITIANE
   COMPATTE E DELLE RELATIVE MOLTEPLICITA'

ROMA - Istituto Matematico dell'Università

<u>C</u>alcolo degli autovalori e delle autosoluzioni

per operatori non autoaggiunti

L. De Vito

All'Istituto Nazionale per le Applicazioni del Calcolo, si sono spes-
so presentati problemi riconducibili alla determinazione di autovalori ed
autosoluzioni di equazioni lineari non autoaggiunte in spazi di Hilbert, cioè
di equazioni che possono scriversi nella forma

$$(1) \qquad\qquad Eu = \lambda u$$

ove E è una trasformazione lineare, definita in una varietà lineare U
di uno spazio Hilbert S complesso completo e separabile, tale che
$E(U) \subset S$ e tale che, inoltre, sia $(Eu, v) \neq (u, Ev)$, $u, v \in U$.

Ogni volta in cui risultavano soddisfatte le seguenti condizioni:
1) $E(U) \equiv S$, 2) <u>insieme</u> Λ, <u>degli autovalori di</u> (1), <u>privo di punti</u>
<u>d'accumulazione al finito,</u>
veniva applicato un metodo di calcolo degli autovalori e delle autosoluzioni,
proposto dal Prof. Picone, che consiste nel considerare il funzionale

$$F(u, \lambda) = \frac{\| Eu - \lambda u \|^2}{\| u \|^2}, \qquad u \in U\text{-}\omega, \ \lambda \text{ numero complesso}$$

(ove ω è lo zero di S), determinarne, per ogni fissato λ, il minimo,
$\mu_n(\lambda)$, nell'insieme U_n di tutti i punti u di $U - \omega$ che sono della
forma

$$u = \sum_{k=1}^{n} c_k u_k$$

L. De Vito

(c_k numeri complessi, $\{u_k\}$ sistema, arbitrariamente prefissato, di punti linearmente indipendenti di $U - \omega$, completo in U),calcolare tutti i punti $\lambda_1^{(n)}, \lambda_2^{(n)}, \ldots, \lambda_{m_n}^{(n)}$ del piano complesso λ , nei quali la funzione $\mu_n(\lambda)$ presenta dei minimi relativi (è subito visto che l'insieme di tali punti non è vuoto e contiene un numero finito di elementi), assumere i numeri $\lambda_1^{(n)}, \lambda_2^{(n)}, \ldots, \lambda_{m_n}^{(n)}$, così costruiti, come approssimazioni n-esime di altrettanti autovalori di (1) ed assumere i punti $u_{\lambda_k^{(n)}}^{(n)}$ di U_n, che rendono minimo $F(u, \lambda_k^{(n)})$ in U_n, come approssimazioni n-esime di autosoluzioni di (1) corrispndenti all'autovalore approssimato da $\lambda_k^{(n)}$ per $n \to +\infty$ [1].

I numerosi esperimenti numerici eseguiti, se da un lato rivelavano sempre la bontà del metodo stesso, nel senso che mostravano come ogni autovalore di (1) venisse approssimato da qualcuno dei numeri $\lambda_k^{(n)}$, dall'altro ponevano in luce il verificarsi di una circostanza che, dal punto di vista pratico, poteva presentare qualche inconveniente: precisamente accadeva che alcuni dei $\lambda_k^{(n)}$, al crescere di n, convergevano verso numeri complessi che non avevano nulla a che fare con gli autovalori di (1). Il verificarsi di questa circostanza può essere controllato, ad esempio, in un caso limite: quello di una trasformazione E per la quale la (1) sia priva di autovalori; così, se si assume come U l'insieme delle funzioni assolutamente continue in $0 \leqslant x \leqslant \pi$, nulle in x = 0 e dotate di derivata prima di quadrato sommabile in $(0, \pi)$, come S lo spazio di Hilbert delle fun-

---

[1] Esposizioni del metodo di Picone sono state fatte da diversi Autori: M. Nasta ("Rend. Acc. Naz. Lincei" 6, XII, 1930), W. Gröbner ("Jahresber.d. Deut. Mathem. Vereinigung", 48, II, 1938), T. Viola ("Rend. di Mat. e delle sue appl." 5, II, 1941), L. Collatz (Eigenwertprobleme und ihre numerische Behandlung, Chelsea Publ. Co. , New York, 1948, pp. 315-316), H. A. Kramers (Die Grundlagen der Quantentheorie - Quantentheorie des elektrons und der Strahlung Hand - und Jahrsbuch Chem. Phys. D Bd. I, Theorien des Aufbaues der Materie I, II, Leipzig, 1938, pp. 200-201.

L, De Vito

zioni di un quadrato sommabile in $(0, \pi)$, e si pone:

$$Eu \equiv \frac{du}{dx} \qquad u \in U \quad , \qquad u_k \equiv \operatorname{sen} kx \qquad k = 1, 2, \ldots$$

si vede che: $m_n = 1$, $\lambda_1^{(n)} = 0$ per ogni n; e, d'altra parte, lo zero non è autovalore per l'equazione $\frac{du}{dx} = \lambda u$ $u \in U$, la quale, anzi, è priva di autovalori. Per così dire, quindi, tra i numeri $\lambda_k^{(n)}$ si osservavano dei valori "parassiti" e nasceva quindi il problema di stabilire un criterio di selezione che permettesse di eliminarli.

Altro problema che veniva posto dall'applicazione del sudetto metodo di Picone, era quello di chiarire in qual modo dovesse intendersi l'approssimazione degli autovalori di (1) da parte dei numeri $\lambda_k^{(n)}$ e quella delle corrispondenti autosoluzioni da parte dei punti $u_{\lambda_k}^{(n)}(n)$. In effetti, il criterio in base al quale, all'Istituto del Calcolo, si sceglieva tra i numeri $\lambda_k^{(n)}$ $(k = 1, 2, \ldots, m_n)$, l' n-esimo approssimante di un dato autovalore di (1), se si rivelava comodo dal punto di vista euristico, non era suscettibile di una giustificazione di carattere generale. Precisamente, si ordinavano i numeri dell'insieme Λ : $\lambda_1, \lambda_2, \ldots$ in successione, e analogamente si ordinavano i $\lambda_1^{(n)}, \lambda_2^{(n)}$, per ogni fissato n, adottando il seguente criterio di ordinamento: se due numeri avevano modulo diverso, si faceva precedere quello di modulo minore, e se due numeri avevano lo stesso modulo, si faceva precedere quello di argomento principale minore; si assumeva, quindi, $\lambda_k^{(n)}$ come approssimazione n-esima di λ_k. A questo proposito, il Professor Fichera osservò che, se E è una trasformazione che possiede due autovalori, uno opposto dell'altro: λ e $-\lambda$, in generale, per una almeno delle due trasformazioni E e $-E$, il procedimento non è valido.

L. De Vito

Un ultimo interrogativo che sorgeva, in relazione al metodo di Picone, era quello di stabilire entro quali ipotesi per la trasformazione E era lecita l'applicazione del metodo stesso; in altre parole, si trattava di fornire una giustificazione teorica di questo metodo di calcolo, entro ipotesi di ragionevole generalità per la E.

A tutti questi interrogativi, venne data esauriente risposta dal Professor Fichera, nel 1955, in una Memoria degli "Annali di Matematica pura e applicata" (vol. XL, serie IV)[2].

L'ipotesi fatta da Fichera sulla trasformazione E è la seguente:

a) E <u>è invertibile e la sua inversa</u> E^{-1} <u>è compatta</u>.

Com'è noto, questa ipotesi è, d'ordinario, verificata quando E sia un operatore differenziale lineare dotato di una funzione di Green (rispetto ad una assegnata condizione al contorno) che possa riguardarsi come nucleo di una trasformazione integrale compatta, ove si assumano convenientemente gli insiemi U ed S (quindi, ad esempio, quando E sia un qualsiasi operatore differenziale lineare ellittico con coefficienti abbastanza regolari).

In tale ipotesi, il Professor Fichera ha dimostrato che, <u>se</u> $\left\{ u_k \right\}$ <u>è un sistema completo in</u> U, <u>tale che</u> $\left\{ E u_k \right\}$ <u>sia completo in</u> S (<u>un sistema siffatto può, ad esempio, costruirsi trasformando, mediante la</u> E^{-1}, <u>un sistema completo in</u> S), <u>fissati comunque due numeri positivi</u> ς <u>ed</u> ε , <u>indicato con</u> $\bigwedge^{(n)}$ <u>l'insieme dei numeri</u> $\lambda_1^{(n)}, \lambda_2^{(n)}, \ldots, \lambda_{m_n}^{(n)}$ <u>relativi al sudetto sistema</u> $\left\{ u_k \right\}$ <u>nel senso sopra specificato, si ha, definitivamente al crescere di</u> n :

---

[2] Dedicata al Professor Picone, in occasione del suo 70-esimo compleanno.

L. De Vito

$$(2) \qquad \Lambda \cap C_\varsigma \subset \mathcal{J}_\varepsilon (\Lambda^{(n)} \cap C_\varsigma)$$

<u>ove</u> C_ς <u>è il cerchio aperto del piano complesso</u> $\lambda : |\lambda| < \varsigma$ e $\mathcal{J}_\varepsilon(\Gamma)$ è l'involucro aperto di raggio ε <u>dell'insieme</u> Γ <u>di tale piano, cioè l'in-</u><u>sieme dei punti che distano</u> da Γ <u>per meno di</u> ε (<u>si conviene che l'insieme</u> <u>vuoto sia contenuto in ogni insieme del piano</u>); in altre parole, nelle dette ipotesi e con le dette assunzioni, fissato comunque un autovalore λ di (1), esiste una successione di numeri $\{\lambda^{(n)}\}$ con $\lambda^{(n)} \in \Lambda^{(n)}$, tale che $\lim_{n \to \infty} \lambda^{(n)} = \lambda$. Inoltre, il Professor Fichera ha mostrato che, indicato con $u^{(m)}_{\lambda^{(m)}}$ un qualunque punto di U_n che renda minimo, in U_n, il funzionale $F(u, \lambda^{(n)})$, la successione di punti $\left\{ u^{(n)}_{\lambda^{(n)}} \| E \, u^{(n)}_{\lambda^{(n)}} \|^{-1} \right\}$ è compatta ed ogni suo elemento di compattezza è un'autosoluzione di (1) relativa all'autovalore $\lambda = \lim_{m \to +\infty} \lambda^{(n)}$ (qui la compattezza è intesa nel senso di Frechet).

Come si è sopra osservato, appunto a causa della presenza di nume-ri "parassiti" tra gli elementi di $\Lambda^{(n)}$, in generale la relazione (2) non può essere invertita, e, in generale, non è vero che, definitivamente al crescere di n, riesca:

$$(3) \qquad \Lambda^{(n)} \cap C_\varsigma \subset \mathcal{J}_\varepsilon (\Lambda \cap \bar{C}_\varsigma),$$

$\bar{C}_\varsigma$ = chiusura di C_ς .

Il Professor Fichera, a questo proposito, ha però dimostrato che <u>le (2) (3) sussistono simultaneamente se</u> $(E^{-1})^* S \equiv U$ [3], <u>purché,</u> <u>in esse, l'insieme</u> $\Lambda^{(n)}$ <u>si sostituisca con il suo sottoinsieme "selezio-</u><u>nato":</u> $L_\varsigma^{(n)}$ <u>costituito da tutti e soli quei</u> $\lambda^{(n)}$ <u>di</u> $\Lambda^{(n)} \cap C_\varsigma$

---

[3] $(E^{-1})^*$ è la trasformazione aggiunta di E^{-1} .

L. De Vito

per i quali riesce:

$$(4) \quad \left[\mathcal{M}_n(\lambda^{(n)})\right]^2 \leqslant \sup_{\lambda \in C_\rho}\left[\left\|(E^* - \bar{\lambda} I)(E - \lambda I) u_\lambda^n\right\|^2 - \mathcal{M}_n(\lambda)^2\right],$$

ove: I è la trasformazione identica, $\bar{\lambda}$ è il coniugato di λ , u_λ^n è un punto di U_n , di norma unitaria, che rende minimo $F(u,\lambda)$ in U_n, $E^* \equiv \left[(E^{-1})^*\right]^{-1}$ (4) è l'aggiunta di E , che riesce definita in U in virtù dell'ipotesi $(E^{-1})^*$ $S \equiv U$, ed ove il sistema $\left\{u_k\right\}$, questa volta, deve godere dell'ulteriore proprietà che $u_k \in (E^{-1})$ U , in modo che abbia senso il calcolo di $(E^* - \bar{\lambda} I) (E - \lambda I)$ nel punto u_λ^n che appartiene a U_n (la possibilità di assoggettare $\left\{u_k\right\}$ a questa ulteriore condizione è bene evidente). Dunque, la (4) fornisce il criterio di selezione cui sopra si accennava; applicando tale selezione ai numeri di $\Lambda^{(n)} \cap C_\rho$, si ottiene un insieme, $L_\rho^{(n)}$, per il quale seguita a sussistere la relazione $\Lambda \cap C_\rho = \mathcal{J}_\varepsilon(L_\rho^{(n)} \cap C_\rho)$ definitivamente al crescere di n , ed inoltre vale anche l'inclusione

$$L_\rho^{(n)} \cap C_\rho \subset \mathcal{J}_\varepsilon(\Lambda \cap \bar{C}_\rho).$$

La dimostrazione con la quale Fichera ha provato tali risultati, fa anche vedere che, nell'ipotesi a), il funzionale $F(u,\lambda)$, per ogni fissato λ , è dotato di minimo in $U - \omega$, ed il valore di tale minimo, $\mu(\lambda)$, è una funzione continua di λ in tutto il piano complesso, che ha come zeri tutti e soli gli autovalori della (1) (la $\mu(\lambda)$ gode, dunque, di alcune delle proprietà della trascendente di Fredholm in relazione ad E^{-1} , ma, a dif-

---

(4) L'invertibilità di $(E^{-1})^*$ è conseguenza di quella di E^{-1} .

L. De Vito

ferenza di questa, non è analitica, anzi, in generale, non è neppure diffe-
renziabile, come si può, ad esempio, constatare nel caso particolare che
E sia autoaggiunta; sotto questa condizione per E , il grafico della
$\mu(\lambda)$, per λ reale, è rappresentato nella figura qui accanto, ove
i λ_k denotano gli autovalori di (1)).

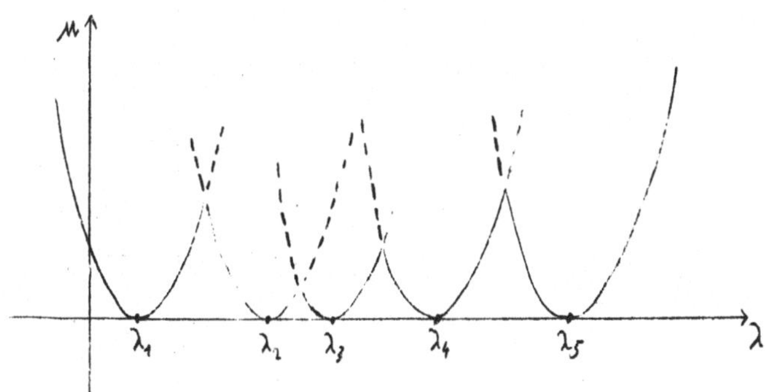

Inoltre, la successione $\left\{\mu_n(\lambda)\right\}$ converge non crescendo verso $\mu(\lambda)$,
uniformemente in ogni insieme limitato del piano.

Procedimenti analoghi a quelli impiegati dal Professor Fichera nella citata Memoria del 1955, hanno consentito ad una sua allieva, la dottores-
sa Bassotti, nel 1961, di provare che il metodo di Picone può anche appli-
carsi al calcolo degli autovalori e delle autosoluzioni di equazioni della forma:

(5) $$\lambda T u = u$$

ove T è una trasformazione lineare compatta di S in sè, esattamente
come nel caso della equazione (1) salvo la sostituzione di U con S ,
del funzionale $F(u,\lambda) = \| Eu - \lambda u\|^2 \|u\|^{-2}$ definito in $U - \omega$ con il
funzionale

L. De Vito

$$F(u, \lambda) = \frac{\|\lambda Tu - u\|^2}{\|u\|^2}$$

definito in $S - \omega$, di $u^{(n)}_{\lambda}(n) \|Eu^{(n)}_{\lambda}(n)\|^{-1}$ con $u^{(n)}_{\lambda}(n) \|u^{(n)}_{\lambda}(n)\|^{-1}$; inoltre, questa volta, $\{u_k\}$ è un qualunque sistema di punti di $S - \omega$ che è assoggettato alla sola condizione di essere completo in S. Anche in questo caso esistono, tra i $\lambda^{(n)}_k$, numeri "parassiti" agli effetti della approssimazione di autovalori della (5); l'insieme selezionato $L^{(n)}_{\varsigma}$ è ora costituito da tutti e soli quei valori $\lambda^{(n)}$ di $\Lambda^{(n)}$ per i quali riesce:

$$\left|\lambda^{(n)}\right| \leqslant \varsigma \quad , \quad \mu_n(\lambda^{(n)}) \leqslant \tfrac{1}{2}$$

$$\left[\mu_n(\lambda^{(n)})\right]^2 \leqslant \sup_{|\lambda| \leqslant \varsigma \, , \, \mu_n(\lambda) \leqslant \frac{1}{2}} \left[\left\|\lambda Tu^n_\lambda + \bar\lambda T^* u^n_\lambda - |\lambda|^2 T^* Tu^n_\lambda\right\|^2 - (1 - \mu_n(\lambda))^2 \right]$$

ove T^* è la trasformazione aggiunta di T, e gli altri simboli hanno lo stesso significato precisato nel caso precedente. [5]

A differenza di quanto accadeva nel caso della (1), l'applicazione del criterio di selezione non richiede alcuna ipotesi aggiuntiva, né sulla trasformazione né sul sistema completo $\{u_k\}$.

L'unica sostanziale differenza tra questo caso e quello precedente è costituita dal fatto che, in generale, lo

$$\inf_{u \in S - \omega} F(u, \lambda) \equiv \mu(\lambda)$$

non è minimo di $F(u, \lambda)$ in $S - \omega$, in corrispondenza ad ogni valore

---

[5] "Atti della Acc. Naz. Lincei", vol. xxx, Maggio, Giugno 1961.

L. De Vito

di λ (come invece succedeva nel caso precedente); in effetti, per un valore di λ tale che $\mu(\lambda) = 1$, può accadere che $F(u,\lambda)$ non sia dotata
di minimo in $S - \omega$, come ad esempio si verifica per le trasformazioni
T , lineari, autoaggiunte, compatte, definite in segno. Si ha però che
$F(u,\lambda)$ è dotata di minimo in $S - \omega$, per ogni valore di λ tale che
sia $\mu(\lambda) < 1$; in questo insieme la $\mu(\lambda)$ gode di tutte le proprietà
che essa verificava nel caso precedente. Se T è autoaggiunta definita positiva, il grafico di $\mu(\lambda)$, per λ reale, è rappresentato nella figura
qui accanto.

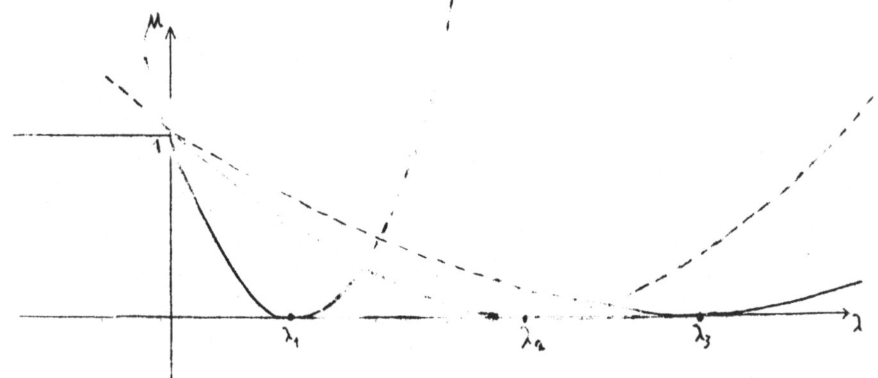

Sul calcolo per difetto e per eccesso degli autovalori delle trasformazioni hermitiane compatte e delle relative molteplicità

L. De Vito

Nelle applicazioni si presenta, sovente, il problema di dover calcolare gli autovalori di una matrice hermitiana sia "per difetto" che "per eccesso", cioè il problema di determinare, in corrispondenza ad ogni autovalore λ della matrice, una successione di numeri convergente per difetto, ed una convergente per eccesso, verso il numero λ .

Sia dunque   A   una matrice hermitiana e siano $\mu_1, \mu_2, \ldots, \mu_r$ i suoi autovalori, disposti in ordine di modulo decrescente e ciascuno ripetuto tante volte quant'è la sua molteplicità. Supponiamo che riesca:

$$\left|\mu_1\right| = \left|\mu_2\right| \ \cdots \ = \left|\mu_{p_1}\right|, \ \left|\mu_{p_1+1}\right| = \left|\mu_{p_1+2}\right| = \cdots = \left|\mu_{p_1+p_2}\right|, \ldots$$

Posto: $\left|\mu_{p_1}\right| = \lambda_1, \left|\mu_{p_1+p_2}\right| = \lambda_2, \ldots, \ A^n \equiv ((a_{hk}^{(n)}))$

si ha, com'è ben noto:

$$\sum_{h,k}^{1,n} \left| a_{hk}^{(n)} \right|^2 = p_1 \lambda_1^{2n} + p_2 \lambda_2^{2n} + \ldots.$$

Se indichiamo con t_n il primo membro di tale relazione, è anche noto che la conoscenza dei numeri t_n consente la determinazione di una successione $\left\{ b_{1,n} \right\}$ (di una successione $\left\{ a_{1,n} \right\}$) approssimante per difetto (per eccesso) il quadrato del più grande autovale, in modulo, di   A : λ_1^2 .

L. De Vito

Basta, per questo, assumere:

$$b_{1,n} = \frac{t_n + 1}{t_n} \quad , \qquad a_{1,n} = \sqrt[n]{t_n} \quad .$$

Analogamente si ha che, posto:

$$b_{2,n} = \frac{t_{n+1} - p_1 \lambda_1^{2n+2}}{t_n - p_1 \lambda_1^{2n}} = \frac{p_2 \lambda_2^{2n+2} + p_3 \lambda_3^{2n+2} + \dots}{p_2 \lambda_2^{2n} + p_3 \lambda_3^{2n} + \dots} \quad ,$$

$$a_{2,n} = \sqrt[n]{t_n - p_1 \lambda_1^{2n}} = \sqrt[n]{p_2 \lambda_2^{2n} + p_3 \lambda_3^{2n} + \dots} \quad ,$$

la successione $\left\{ b_{2,n} \right\}$ (la successione $\left\{ a_{2,n} \right\}$) converge per difetto
(per eccesso) verso λ_2^2 , cioè verso il quadrato del secondo autovalore,
in modulo, di A . In modo analogo si costruiscono le successioni approssimanti per difetto e per eccesso i numeri λ_3^2 , λ_4^2 ,... .

Tuttavia, mentre il risultato relativo alle due successioni $\left\{ b_{1,n} \right\}$,
$\left\{ a_{1,n} \right\}$ è utilizzabile praticamente, per il calcolo effettivo di λ_1^2 ,
potendosi calcolare numericamente i t_n in numero comunque grande, il
risultato relativo alle due successioni $\left\{ b_{2,n} \right\}$, $\left\{ a_{2,n} \right\}$ presenta solo un
interesse teorico, poichè in pratica, non si dispone né della conoscenza di
p_1 né di quella -esatta- di λ_1^2 . Peraltro, se si ammette di conoscere
il numero intero p_1 , ed i termini delle successioni $\left\{ a_{1,n} \right\}$, $\left\{ b_{1,n} \right\}$,
si possono subito stabilire delle limitazioni inferiori e superiori per λ_2^2 ;
si ha infatti:

L. De Vito

$$\beta_{2,n} = \max \left[\ 0,\ \frac{t_{n+1} - p_1\, a_{1,\,m_{n+1}}^{\,n+1}}{t_n - p_1\, b_{1,\,m_n}^{\,n}}\ \right] \leq \lambda_2^{\,2}$$

$$\alpha_{2,n} = (\ t_n - p_1\, b_{1,\,m_n}^{\,n}\ )^{1/n} \geqslant \lambda_2^{\,2}$$

ove $\left\{ m_n \right\}$ è una arbitraria successione crescente di numeri naturali.

A questo punto, viene spontaneo di chiedersi se non si possa appro-fittare della arbitrarietà di $\left\{ m_n \right\}$ per fare in modo che $\left\{ \beta_{2,n} \right\}$ e $\left\{ \alpha_{2,n} \right\}$ convergano verso $\lambda_2^{\,2}$, e possano quindi essere assunte come successio-ni approssimanti in quadrato di λ_2 per difetto e per eccesso rispettiva-mente. A questo interrogativo può darsi risposta positiva, come è mostra-to in una Nota dello scrivente [1], facendo $m_n = n^2$ o, più in generale, assumendo m_n in guisa tale che $\lim\limits_{n \to \infty} \dfrac{m_n}{n} = \infty$, pur di modificare lie-vemente la definizione di $\beta_{2,n}$ e, precisamente, porre

$$\beta_{2,n} = \max \left[\ 0,\ \frac{t_{n-1} - p_1^{\,n/n+1}\ a_{1,\,m_{n+1}}^{\,n+1}}{t_n - p_1\, b_{1,\,m_n}^{\,n}}\ \right].$$

I caratteri di tali convergenze sono messi in evidenza dalle seguenti relazio-ni :

$$\lim_{n \to \infty} (\ \alpha_{k,\,m_n}^{\,n} - p_k\, \lambda_k^{\,2n}\ )^{1/n} = 0\ ,\quad \lim_{n \to \infty} (\ \alpha_{k,\,m_n}^{\,1/n} - p_k^{\,1/n}\, \lambda_k^{\,2}\ )^{1/n} = 0,$$

____

[1] "Sul calcolo approssimato degli autovalori delle trasformazioni compat-te e delle relative molteplicità" Rend. Acc. Naz. Lincei, xxx, 1961.

L. De Vito

$$\lim_{n \to \infty} (\lambda_k^{2n} - \beta_{k,m_n}^{n})^{1/n} = 0 \, , \quad \lim_{n \to \infty} (\lambda_k^{2} - \beta_{k,m_n})^{1/n} = 0,$$

$$k = 1, 2 \, , \quad \alpha_{1,s} = a_{1,s} \quad \beta_{1,s} = b_{1,s}$$

Considerazioni perfettamente analoghe a queste possono ripetersi in relazione a λ_3^2, λ_4^2 etc...., pur di conoscere, di volta in volta, rispettivamente, i numeri p_1, p_2, i numeri p_1, p_2, p_3 etc....

Per quanto riguarda la vautazione dei numeri p_k, non si conoscono ancora procedimenti di carattere generale; si può solo dire che la successione $\{ r_{1,n} \}$, con

$$r_{1,n} = \frac{t_n}{b_{1,m_n}^{n}} \, ,$$

approssima per eccesso p_1, la successione $\{ r_{2,n} \}$, con

$$r_{2,n} = \frac{t_n - p_1 \, b_{1,m_n}^{n}}{\beta_{2,m_n}^{n}} \, ,$$

approssima per eccesso p_2, etc...., ove $\{ m_n \}$ sia scelto nello stesso modo sopra indicato (cifr. loc. cit. in [1] pag. 456). Si ha, infine, che la conoscenza di un numero separatore della coppia λ_1, λ_2, cioè un numero λ tale che $\lambda_2 \leqslant \lambda < \lambda_1$, è perfettamente equivalente alla conoscenza di p_1. E', intanto, ben evidente, in base a quanto sopra detto, che, dalla conoscenza di p_1, si deduce quella di λ; viceversa, assunto $\bar{n}$ tale che $b_{1,\bar{n}} > \lambda^2$, ε tale che $0 < \varepsilon < 1$, $\varepsilon < \frac{1}{2} (b_{1,\bar{n}} - \lambda^2)$ ed n_0 tale che

L. De Vito

$$b_{1,m_{n_0}} > 0 \quad , \qquad \frac{a_{1,m_{n_0}}}{b_{1,m_{n_0}}} - 1 < \frac{\varepsilon}{3r_{1,m_{n_0}}}$$

$$a_{1,m_{n_0}} - b_{1,m_{n_0}} < \min\left[\varepsilon \ , \ \frac{\varepsilon}{2} \ \frac{b_{1,\bar{n}} - \lambda^2}{r_{1,n_0+1}}\right]$$

(si può **vedere** che un indice siffatto certamente esiste) si ha:

$$p_1 = \text{parte intera di } r_{1,n_0}$$

(cfr. loc. cit. in [1] pag. 457).

Le considerazioni testè svolte in relazione al calcolo degli autovalori di una matrice hermitiana, possono ripetersi, senza alcuna modificazione
formale, con riguardo al calcolo degli autovalori di una trasformazione lineare **T** di uno spazio di Hilbert S (complesso completo e separabile)
in sè, la quale sia hermitiana, compatta e tale inoltre che le serie

$$\sum_k p_k \lambda_k^{2n}$$

(ove p_k e λ_k hanno lo stesso significato precisato sopra) risultino con-
vergenti, almeno da un certo n in poi, e le loro rispettive somme siano
numericamente determinabili. E' questo, sostanzialmente, il caso delle
trasformazioni hermitiane integrali $T(u) = \int_D K(x, y) u(y)dy, \left[K(x,y) = \overline{K(y,x)}\right]$ dello spazio di Hilbert $\mathcal{L}^2(D)$ in sé, aventi un nucleo $K(x,y)$
tale che uno almeno dei suoi iterati sia di quadrato sommabile in $D \times D$.

Le idee ora esposte, possono servire, tra l'altro, per integrare il
classico metodo di <u>Ritz</u> per il calcolo degli autovalori. Si consideri infat-

L. De Vito

ti, ad esempio, la trasformazione integrale T dianzi introdotta, con il
nucleo K verificante le condizioni già menzionate ed inoltre definito posi-
tivo; ove non si conosca la molteplicità di λ_1 , il metodo di Ritz non consente di limitare inferiormente il numero λ_2 . Se però si sa, applicando
i procedimenti sopra descritti, che il numero intero m - 1 limita supe-
riormente la molteplicità di λ_1 , si può senz'altro assumere, come limi-
tazione inferiore per λ_2 , l' m-esima radice dell'equazione

$$\det \left(\left(\int_D T(w_i)\,\overline{w_{\ell}}\ dy - \lambda \int_D w_i\,\overline{w_{\ell}}\ dy \right)\right) = 0$$

(ove w_1, w_2,... son funzioni linearmente indipendenti).

Un'altra possibilità di integrare il metodo di Ritz (per trasformazioni integrali aventi il nucleo verificante le dette proprietà) è fornita dalla se-
guente osservazione. Se le lettere $p_1^{(m)}$, $p_2^{(m)}$,...., $\lambda_1^{(m)}$, $\lambda_2^{(m)}$,....,
$t_n^{(m)}$ hanno significato analogo a quello delle p_k, λ_k, t_n in relazione
alla m-esima trasformazione approssimante di Ritz, si ha :

$$0 \leqslant \lambda_k^{2n} - \left|\lambda_k^{(m)}\right|^{2n} \leqslant t_n - t_n^{(m)}, \quad \lim_{m\to\infty}\left[t_n - t_n^{(m)}\right] = 0$$

come è immediato constatare. Ne viene che, accanto alla successione
$\left\{\lambda_k^{(m)}\right\}_m$ fornita dal metodo di Ritz, che converge a λ_k per difetto, si può considerare anche la successione $\left\{\nu_k^{(m)}\right\}_m$ con

$$\nu_k^{(m)} = \left(\left|\lambda_k^{(m)}\right|^{2n} - t_n^{(m)} + t_n \right)^{1/2n}$$

che converge a λ_k per eccesso.

CENTRO INTERNAZIONALE MATEMATICO ESTIVO

(C. I. M. E.)

J. B. DIAZ

UPPER AND LOWER BOUNDS FOR THE

TORSIONAL RIGIDITY AND THE CAPACITY,

DERIVED FROM THE INEQUALITY OF SCHWARZ

ROMA, Istituto Matematico dell'Università

UPPER AND LOWER BOUNDS FOR THE TORSIONAL RIGIDITY

AND THE CAPACITY, DERIVED FROM THE INEQUALITY OF

SCHWARZ

by J.B. DIAZ

(Institute for Fluid Dynamics and Applied Mathematics, University of Maryland)

1. Introduction.

In many problems of mathematical physics it is desired to find the numerical value of a quadratic integral of an unknown function, where the unknown function is a solution of a linear boundary value problem consisting of a linear partial differential equation plus linear boundary condition. The quadratic integral in question is usually the quadratic form occurring in a Green's identity for the differential operator involved in the boundary value problem. The present exposition is concerned with two particular instances of this general situation. In section 2, which is based upon references $[6]$ and $[7]$ in the bibliography, upper and lower bounds for the torsional rigidity of a cylindrical beam are derived from Schwarz's inequality.

Section 3 is devoted to the estimation of the capacity, and is based upon references $[1]$ and $[13]$ in the bibliography.

The brief bibliography contains references in which a fuller discussion of the topics mentioned is to be found, and is by no means meant to be exhaustive.

2. Upper and lower bounds for the Dirichlet integral.

Let p, q, P, Q be sufficiently smooth real valued functions defined on D + C , where D is a bounded plane domain with a (smooth) boundary C. For α any real number, one has that

J. B. Diaz

$$\int_D \left[(p + \alpha P)^2 + (q + \alpha Q)^2 \right] \, dxdy \geqslant 0 \; ,$$

from which it follows that

$$\left(\int_D (pP + qQ)dxdy \right)^2 \leqslant \int_D (p^2 + q^2)dxdy \int_D (P^2 + Q^2)dxdy. \quad (S)$$

This last inequality, which will be referred to as Schwarz's inequality, is the starting point for all the upper and lower bounds for the Dirichlet integral to be given here.

Consider the determination of upper and lower bounds for the Dirichlet integral

$$\int_D (v_x^2 + v_y^2)dxdy$$

of a solution $v(x,y)$ of Neumann's problem

$$\Delta v = v_{xx} + v_{yy} = 0 \; , \qquad \text{on } D \; ,$$

$$\frac{\partial v}{\partial n} = v_x n_x + v_y n_y = f \; , \text{ on } C \; .$$

(Here, n_x and n_y denote the components of the <u>outer</u> unit normal to the boundary C). The desired bounds follow at once upon choosing suitably the functions p, q, P, Q which appear in the Schwarz's inequality (S).

To derive an upper bound, let p, q be such that

$$p_x + q_y = 0 \; , \qquad\qquad \text{on } D \; ,$$

$$p n_x + q n_y = \frac{\partial v}{\partial n} = f \; , \quad \text{on } C \; ,$$

and let P, Q be given by

$$P = v_x \; , \qquad Q = v_y \; .$$

Then, by integration by parts, and Green's identity, one readily obtains

J. B. Diaz

$$\int_D (pP + qQ)dxdy = \int_D (pv_x + qv_y)dxdy$$

$$= -\int_D v(p_x + q_y)dxdy + \int_C v(pn_x + qn_y)ds$$

$$= \int_C v\,\frac{\partial v}{\partial n}\,ds = \int_D (v_x^2 + v_y^2)dxdy \ ;$$

which, together with (S), yields

$$\int_D (v_x^2 + v_y^2)dxdy \ \leqslant \ \int_D (p^2 + q^2)dxdy \ .$$

(This inequality is precisely Kelvin's minimum kinetic energy theorem, see Lamb $\big[$11, pages 47 and 57$\big]$, and also Diaz and Weinstein $\big[$7, page 109$\big]$). Since the condition $p_x + q_y = 0$ on D can always be replaced by $p = u_y$ and $q = -u_x$, where u is a suitable function (not necessarily single-valued), the last inequality may be restated as follows : If u is any function (not necessarily single-valued, but such that u_x and u_y are single-valued) such that

$$\frac{\partial u}{\partial s} = u_x\,\frac{dx}{ds} + u_y\,\frac{dy}{ds} = -u_x n_y + u_y n_x = \frac{\partial v}{\partial n} \ , \quad \text{on C} \ ,$$

then

$$\int_D (v_x^2 + v_y^2)dxdy \ \leqslant \ \int_D (u_x^2 + u_y^2)dxdy \ .$$

A lower bound for the Dirichlet integral of v can be derived similarly. This time, let

$$p = w_x \ , \qquad q = w_y \ ,$$

with $w(x, y)$ a non-constant real valued function defined on $D + C$, and also

$$P = v_x \ , \qquad Q = v_y \ ,$$

as before. Then, by Green's identity,

$$\int_D (pP + qQ)dxdy = \int_D (v_x w_x + v_y w_y)dxdy$$

$$= -\int_D w\,\Delta v\,dxdy + \int_C w\,\frac{\partial v}{\partial n}\,ds$$

$$= \int_C w\,\frac{\partial v}{\partial n}\,ds \ ;$$

which, together with Schwarz's inequality (S), yields

$$\frac{\left(\int_C w \, \frac{\partial v}{\partial n} \, ds\right)^2}{\int_D (w_x^2 + w_y^2)dxdy} \ \leqslant \ \int_D (v_x^2 + v_y^2)dxdy \ .$$

Combining the two inequalities already obtained, one may summarize the results thus :

$$\frac{\left(\int_C w \, \frac{\partial v}{\partial n} \, ds\right)^2}{\int_D (w_x^2 + w_y^2)dxdy} \ \leqslant \ \int_D (v_x^2 + v_y^2)dxdy \ \leqslant \ \int_D (u_x^2 + u_y^2)dxdy \ ,$$

that is to say, any non-constant function w furnishes a lower bounds, and any function u satisfying the boundary condition $\frac{\partial u}{\partial s} = \frac{\partial v}{\partial n}$ on C furnishes an upper bound, for the Dirichlet integral of a solution v of Neumann's problem.

(While it will not be derived here in detail, the derivation being similar to that just carried out in the case of the Neumann boundary value problem, it will be remarked that a corresponding result is valid for a solution v of the Dirichlet boundary value problem

$$\Delta v = v_{xx} + v_{yy} = 0 \ , \qquad \text{on } D \ ,$$

$$v = f \qquad , \qquad \text{on } C \ .$$

In this instance, one has

$$\frac{\left(\int_C v \, \frac{\partial w}{\partial s} \, ds\right)^2}{\int_D (w_x^2 + w_y^2)dxdy} \ \leqslant \ \int_D (v_x^2 + v_y^2)dxdy \ \leqslant \ \int_D (u_x^2 + u_y^2)dxdy \ ;$$

that is to say, any non-constant function w furnishes a lower bound, and any function u satisfying the boundary condition u = v on C furnishes an upper bound, for the Dirichlet integral of a solution v of Dirichlet's problem. The right hand inequality is nothing else but Dirichlet's principle, while the left hand i-

J.B.Diaz

nequality contains as a special case a lower bound for the Dirichlet integral of a solution of Dirichlet's problem, given by Trefftz $[16]$, in terms of an arbitrary non-constant harmonic function.)

Only a single example of the many possible applications of the upper and lower bounds given for the Dirichlet integral of a solution of Neumann's problem will be indicated. Consider the torsion of an elastic cylindrical beam of cross section D ; and assume that Lame's constant of elasticity, μ , is taken to be unity. The stiffness, or torsional rigidity, S, of such a cylindrical beam, is given by the formula (Diaz and Weinstein $[7$, p. $108]$)

$$S = P - \int_D (v_x^2 + v_y^2)dxdy \; ,$$

where P is the polar moment of inertia of the domain with respect to its centroid, and v is the warping function, which is a solution of the Neumann boundary value problem

$$\Delta v = v_{xx} + v_{yy} = 0 \; , \qquad \text{on D} \; ,$$

$$\frac{\partial v}{\partial n} = \frac{\partial}{\partial s}\left[\frac{1}{2}(x^2 + y^2) \right] \; , \qquad \text{on} \quad C \; ,$$

with $\dfrac{\partial}{\partial s}$ denoting differentiation along the boundary C .

Parenthetically, notice that this formula for the torsional rigidity implies that for any (simple or multiply) connected section one has that

$$S \leqslant P \; ,$$

with equality if and only if the Dirichlet integral of the warping function is zero, which means that the warping function must be a constant in view of the boundary condition satisfied by the function v , this can only happen if the domain D is either a circle or a circular ring. Curiously, therefore, for domains whose connectivity is more than three, the torsional rigidity is al-

J. B. Diaz

ways less than the polar moment of inertia with respect to the centroid. An alternative way of looking at this is the following "isoinertial" principle : of all domains with prescribed polar moment of inertia with respect to the centroid, the circle and the circular ring possess the maximum torsional rigidity.

3. Upper and lower bounds for the capacity.

The capacity C of a smooth surface ("conductor") S in three dimensional space may be taken to be defined by the equation

$$C = \frac{1}{4\pi} \int_D \left|\operatorname{grad} v\right|^2 dxdydz \ ,$$

where D is the region exterior to the surface S , and v is the solution of the exterior Dirichlet problem

$$\Delta v = v_{xx} + v_{yy} + v_{zz} = 0 \ , \qquad \text{on} \ \ D \ ,$$

$$v = 1 \qquad\qquad\qquad , \qquad \text{on} \ \ S \ ,$$

$$\lim_{(x,y,z) \to \infty} v(x,y,z) = 1 \ .$$

From Dirichlet's principle, an upper bound for the capacity C is given by

$$C \leqslant \frac{1}{4\pi} \int_D \left|\operatorname{grad} w\right|^2 dx\, dy\, dz \ ,$$

where $w(x,y,z)$ is a sufficiently smooth function (i.e., continuous, with piecewise continuous first partial derivatives) such that $w(x,y,z) = 1$ if (x,y,z) is a point of S and also that $w = O(r^{-1})$ as $r = (x^2 + y^2 + z^2)^{1/2}$ approaches infinity.

The purpose of this section is to show how, using a simple "trial function" w (which seems to be naturally dictated by the symmetry of the domains in question) it is possible to obtain fairly close, readily computable

J. B. Diaz

upper bounds for the capacity of the regular solids. For definiteness in describing the procedure, let S be the cube, of side 2 and of vertice $(\pm 1, \pm 1, \pm 1)$, which is circumscribed about the unit sphere with center at the origin.

The exterior D of the cube S is divided into six congruent infinite pyramidal domains (one corresponding to each face of the cube). The trial function w to be chosen will first be defined on one such (frustrated) pyramidal domain, and then the definition of the function will be extended, "by symmetry" over the rest of the exterior, D .

Consider the pyramidal frustrum D' (corresponding to the face of the cube which contains the point $(1,0,0)$) consisting of the set of points (x,y,z) satisfying the three inequalities $x \geqslant 1$, $|y| \leqslant x$, $|z| \leqslant x$, and suppose that the trial function has been defined on D', and then the definition of w has been extended symmetrically to the entire exterior of the cube, as explained above.

Then

$$\int_D |\text{grad } w|^2 \, dx \, dy \, dz = 5 \int_{D'} |\text{grad } w|^2 \, dx \, dy \, dz \; ;$$

and if, in particular, one sets $w(x,y,z) = f(x)$ on D', it follows that

$$\int_D |\text{grad } w|^2 \, dx \, dy \, dz =: 6 \int_1^\infty (4x^2) \left[f'(x) \right]^2 \, dx \; .$$

"Minimizing" this last integral (with respect to all admissible functions f) leads to the Euler-Lagrange equation

$$\frac{\partial}{\partial f} \left\{ x^2 \left[f'(x) \right]^2 \right\} - \frac{d}{dx} \frac{\partial}{\partial f'} \left\{ x^2 \left[f'(x) \right]^2 \right\} = 0 \; ,$$

that is

$$2x \, f'(x) + x^2 \, f''(x) = 0 \; ;$$

J. B. Diaz

which, since $f(1) = 1$ and $f(x) = O(x^{-1})$ for large x, means that

$$f(x) = \frac{1}{x}$$

gives the "best" possible choice of the function f. This very particular choice of the function f already leads to a fairly good upper bound for the capacity C of a cube, since then

$$-\frac{1}{4\pi} \int_D |\operatorname{grad} w|^2 \, dx \, dy \, dz = -\frac{6}{4\pi} \int_1^\infty (4x^2) \left[\frac{d}{dx} \frac{1}{x} \right]^2 dx = \frac{6}{\pi},$$

and it is known (see below) that the capacity of a cube is of the order of 1.2. (Notice that the simple trial function just described may be used "a priori" entirely independently of the possible justification of the heuristic variational consideration which led to its discovery.)

An improved upper bound for the capacity C may be obtained by letting

$$w(x, y, z) = \frac{1}{x} + (\lambda |y| + \mu z) \left(\frac{1}{x^3} - \frac{1}{x^2} \right)$$

on D', where λ and μ are real numbers, and minimizing the resulting Dirichlet integral with respect to λ and μ. Notice that the trial function w defined in D by symmetry satisfies the boundary condition $w = 1$ on the cube, since the additional terms are just "coordinate functions" (in the terminology of Walther Ritz). The final result of the computation, which will be omitted here, is that (for a cube of side 2) one has $C \leqslant 1.6103\ldots$ Using the volume-radius, i. e. the radius of the sphere of the same volume as the the interior volume of the cube (see Polya-Szegö $[13, \text{p. }23]$), gives the lower bound $1.240 \leqslant C$. The whole "symmetry" process described above for the cube is carried out in full, for any regular solid, in J. Conlan, J. B. Diaz, and W. E. Parr $[1]$. For the capacity C of the icosahedron circumscribed about the unit sphere, the upper bound obtained there is that

J. B. Diaz

C $\leqslant$ 1.096, while the lower bound obtainable from the volume-radius is

1.064 $\leqslant$ C . The other regular solids can be treated similarly.

At the 1954 conference at the University of Trieste, it was reported that (see Diaz $\left[4\right]$) for the capacity C of a cube of edge 2 one has

$$1.308 \;<\; C \;<\; 1.336 \quad,$$

where the lower bound was given by Daboni $\left[8\right]$ and the upper bound was given by Payne and Wamberger $\left[16\right]$.

Quite recently, W. E. Parr $\left[18\right]$, obtained the upper bound 1.335, as an application of his extension of Polya-Szegö's $\left[13\right]$ method of prescribed level surfaces.

J. B. Diaz

BIBLIOGRAPHY

1. James Conlan, J. B. Diaz and W. E. Parr, On the capacity of the icosahedron, Journal of Mathematics and Physics, vol. 2, 1961, 259-261.

2. R. Courant and D. H. Hilbert, Methods of mathematical physics, First English edition, New York, 1953.

3. J. B. Diaz, Upper and lower bounds for quadratic functionals, Collectanea Mathematica, Seminario Matematico de Barcelona, vol. 4, 1951, 3-50.

4. J. B. Diaz, Some recent results in linear partial differential equations, Atti del convegno internazionale sulle equazioni alle derivate parziali, Trieste, 1954, Edizioni Cremonese, Roma, 1955, 1-29.

5. J. B. Diaz, Upper and lower bounds for quadratic integrals, and at a point, for solutions of linear boundary value problems, in Boundary Problems in Differential Equations, edited by Rudolph E. Langer, The University of Wisconsin Press, Madison, 1960, 47-83.

6. J. B. Diaz and A. Weinstein, Schwarz's inequality and the methods of Rayleigh-Ritz and Trefftz, Journal of Mathematics and Physics, vol. 26, 1947, 133-136.

7. J. B. Diaz and A. Weinstein, The torsional rigidity and variational methods, American Journal of Mathematics, vol. 70, 1948, 107-116.

J. B. Diaz

8. L. Daboni, Applicazione al caso del cubo di un metodo per il calcolo per
    eccesso e per difetto della capacità elettrostatica di un conduttore, Rend.
    Acc. Naz. Lincei, sez. VIII, vol. XIV, 1953, 461-466.

9. G. Fichera, Risultati concernenti la risoluzione delle equazioni funzionali
    dovuti all'Istituto Nazionale per le applicazioni del Calcolo, Memorie del-
    l'Accademia Nazionale dei Lincei, serie VIII, vol. 3, 1950, pp. 59-68.

10. G. Fichera, Methods of linear functional analysis in mathematical physics,
    Proceedings of the International Congress of Mathematicians, Amsterdam,
    1954, vol. III, 216-228.

11. H. Lamb, Hydrodynamics, Sixth edition, New York, 1945.

12. M. Picone - G. Fichera, Neue funktionalanalytische Grundlage für die Exi-
    stenzprobleme und Lösungamethoden von Systemen linear partieller Diffe-
    rentialgleichungen, Monatshefte für Mathematick, vol. 54, 1950, 188-209.

13. G. Pólya and G. Szegö, Isoperimetric inequalities in mathematical physics,
    Princeton University Press, 1951 (see also the book review in Bulletin of
    the American Mathematical Society, vol. 59, 1953, pp. 588-602).

14. W. Prager and J. L. Synge, Approximations in elasticity based on the con-
    cept of function space, Quarterly of Applied Mathematics, vol. 5, 1947,
    241-269.

15. J. L. Synge, The hypercircle in mathematical physics, Cambridge Univer-
    sity Press, 1957 (see also the book review in Bulletin of the American Ma-
    thematical Society, vol. 65, 1959).

16. <u>L. E. Payne and H. F. Weinberger</u>, New bounds in harmonic and biharmonic problems, Journal of Mathematics and Physics, vol. 33, 1955, 291-307.

17. <u>E. Trefftz</u>, Ein Gegenstück zum Ritzchen Verfahren, Proc. Second International Congress Applied Mechanics, Zürich, 1927, 131-137.

18. <u>A. Weinstein</u>, New methods for the estimation of the torsional rigidity, Proceedings of the Third Symposium in Applied Mathematics, American Mathematical Society, 1950, 141-161.

19. <u>W. E. Parr</u>, Upper and lower bounds for the capacitance of the regular solids, Journal of the Society for Industrial and Applied Mathematics, vol. 9, 1961, 334-386.

CENTRO INTERNAZIONALE MATEMATICO ESTIVO

(C. I. M. E.)

MENAHEM SCHIFFER

FREDHOLM EIGENVALUES AND CONFORMAL MAPPING

Roma - Istituto Matematico dell'Università

FREDHOLM EIGENVALUES AND CONFORMAL MAPPING

by Menahem Schiffer

Introduction

This paper is mainly of expository nature. Its aim is to give a connected account of various results regarding the Fredholm eigenfunctions and the Fredholm eigenvalues of plane domains. The Fredholm eigenvalues of a curve system are a set of functionals of these curves whose study leads to may useful applications. They are closely related to the boundary value problem for harmonic functions. They are important in the theory of conformal mapping, in the theory of kernel functions and orthonormal series. They play a role in the theory of the Hilbert transform and in the theory of univalent functions. Their dependence on the curve system is displayed by a very elegant and convenient variational formula. Some applications of these results are given to show the significance of the identities and formulas obtained. For a more detailed development of many points the reader is referred to $[10]$, $[11]$ and $[12]$.

1. Boundary Value Problem of Potential Theory and Fredholm Eigenvalues

Let C be a closed curve in the complex z-plane with interior domain D and exterior domain $\tilde{D}$. We suppose C to be three times continuously differentiable. It possesses at every point $\zeta \in C$ a normal n_ζ , and it is well known that the kernel

$$(1) \qquad k(z, \zeta) = \frac{\partial}{\partial n_\zeta} \log \frac{1}{|z - \zeta|}$$

is a continuous function of both argument points z and ζ on C . We shall

M. Schiffer

always assume the normal n_ζ to be directed into D.

The kernel $k(z, \zeta)$ was introduced by Poincaré in order to solve the first boundary value problem of potential theory. Suppose that we are given a continuous function $f(z)$ on the curve C and wish to find a function $u(z)$ harmonic in D which takes on C the boundary values $f(z)$. We try to solve this problem by superposition of elementary harmonic functions and make the "Ansatz"

$$(2) \qquad u(z) = \frac{1}{\pi} \oint_C k(z, \zeta)\, \varphi(\zeta)\, ds_\zeta$$

with a weight function $\varphi(\zeta)$ still to be determined for our problem. This is clearly a harmonic function for $z \in D$. As $z \to z_0 \in C$, we have the well-known jump condition

$$(3) \qquad \lim_{z \to z_0} u(z) = \varphi(z_0) + \frac{1}{\pi} \oint_C k(z_0, \zeta)\, \varphi(\zeta)\, ds_\zeta$$

Since we have prescribed the boundary values $f(z)$ of $u(z)$, we obtain for the determination of the weight function $\varphi(\zeta)$ the Fredholm integral equation of the second kind

$$(4) \qquad f(z) = \varphi(z) + \frac{1}{\pi} \oint_C k(z, \zeta)\, \varphi(\zeta)\, ds_\zeta$$

This particular inhomogeneous integral equation with continuous kernel established by Poincaré led Fredholm to his general theory of linear integral equations. He established his celebrated alternative :

Either the integral equation (4) possesses a nontrivial solution $\varphi(z)$ for $f \equiv 0$, or it possesses a unique solution $\varphi(z)$ to each given integrable function $f(z)$.

M. Schiffer

Thus, the entire boundary value theory of harmonic functions in the plane is reduced to the study of the homogeneous integral equation

$$(5) \qquad \varphi_\nu(z) = \frac{\lambda_\nu}{\pi} \oint_C k(z, \zeta) \, \varphi_\nu(\zeta) ds_\zeta$$

If it could be shown that -1 is not an eigenvalue λ_ν of (5), the unique existence of a harmonic $u(z)$ with prescribed boundary values $f(z)$ on C would have been established. This was indeed done by Fredholm and the existence problem of the function $u(z)$ was solved.

There remains the interesting question to study the eigenvalues λ_ν connected with this important integral equation and to determine their potential theoretical significance. They are called the Fredholm eigenvalues of the curve C ; they may also be considered as functionals of the domain D or of the domain $\tilde{D}$.

It is well known that the lowest eigenvalue (in absolute value) λ_0 has the value $+1$ and belongs to the eigenfunction $\varphi_0(\zeta) = \text{const.}$ All other eigenvalues are real and satisfy $|\lambda_\nu| > 1$. Since we easily can handle the contribution of the trivial eigenvalue λ_0 to the kernel $k(z, \zeta)$, we actually can reduce the numerical solution of the first and second boundary value problem of potential theory to the process of iteration, i. e. , the Liouville-Neumann series development. The details of this procedure were developed by Gershgorin [6] . It probably is the most convenient numerical method in two-dimensional potential theory.

However, the speed of convergence of the Liouville-Neumann series depends on the absolute value of the lowest nontrivial Fredholm eigenvalue [1, 9, 17] . Thus, we see the significance to estimate the nontrivial Fredholm eigenvalues of a given curve C . We propose to study the values $\lambda_\nu(C)$ as

M. Schiffer

functionals or as "fonction de ligne" in the sense of Volterra and to solve extremum problems concerning them.

2. The Fredholm Eigenfunctions

We consider the Fredholm eigenfunctions $\varphi_\nu(\zeta)$ connected with the eigenvalues λ_ν. These functions are defined only on the curve C. We now introduce harmonic functions $h_\nu(z)$ and $\tilde{h}_\nu(z)$ defined in D and $\tilde{D}$, respectively, which are closely related to the $\varphi_\nu(\zeta)$.

We define

$$(6) \qquad h_\nu(z) = \frac{\lambda_\nu}{\pi} \oint_C k(z,\zeta)\, \varphi_\nu(\zeta)\,ds_\zeta$$

The right-hand side of (6) represents a harmonic function of z if z ranges over D; it also represents a harmonic function if z varies in $\tilde{D}$. But this harmonic function is not the continuation of $h(z)$ defined in D. For the sake of clarity we therefore shall denote it by the letters $\tilde{h}(z)$.

The jump conditions for the kernel $k(z,\zeta)$ imply for each point $z_0 \epsilon C$

$$(7) \qquad \lim_{z \to z_0} h_\nu(z) = (1+\lambda_\nu)\,\varphi_\nu(z_0)\,, \qquad \lim_{z \to z_0} \tilde{h}_\nu(z) = (1-\lambda_\nu)\,\varphi_\nu(z_0)$$

These boundary conditions on C determine the harmonic functions $h_\nu(z)$ and $\tilde{h}_\nu(z)$ uniquely. It is also well known from the theory of the double-layer potential that everywhere on C

$$(8) \qquad \frac{\partial\, h_\nu(z_0)}{\partial n} = \frac{\partial\, \tilde{h}_\nu(z_0)}{\partial n}$$

M. Schiffer

We have thus solved the following interesting problem of potential theory :
To determine a harmonic function $h(z)$ in D and a harmonic function $\tilde{h}(z)$
in $\tilde{D}$ with the same normal derivatives at the common boundary and having
proportional boundary values there.

We can also give another interpretation to this result. Let

$$(9) \qquad D[h] = \iint_D (\nabla h)^2 \, d\tau = - \oint_C h \frac{\partial h}{\partial n} \, ds$$

$$D[\tilde{h}] = \iint_{\tilde{D}} (\nabla \tilde{h})^2 \, d\tau = + \oint_C \tilde{h} \frac{\partial \tilde{h}}{\partial n} \, ds$$

denote the Dirichlet integrals of h and $\tilde{h}$. It follows from (7) and (8) that

$$(10) \qquad D[h_\nu] = \frac{\lambda_\nu + 1}{\lambda_\nu - 1} \cdot D[\tilde{h}_\nu]$$

Since Dirichlet integrals are obviously nonnegative; we read off from (10)
that $|\lambda_\nu| \geq 1$ and that $|\lambda_\nu| = 1$ is only possible for $\varphi_\nu(\zeta) = \text{const.}$
Finally, it is easy to see that $\varphi_\nu = \text{const}$ implies $\lambda_\nu = +1$. Thus, the i-
dentities (7) and (8) decide the essential step in the above mentioned Fredholm
alternative.

Consider next an arbitrary continuous function $f(z)$ defined on C ; determine the harmonic functions $h(z)$ and $\tilde{h}(z)$ in D and $\tilde{D}$ which have the
common boundary values $f(z)$ on C. Let $D[h]$ and $D[\tilde{h}]$ be their Dirichlet integrals and consider the ratio $D[h]/D[\tilde{h}]$ as a functional of $f(z)$.
It is easily seen that for all f holds

$$(11) \qquad \frac{\lambda_1 - 1}{\lambda_1 + 1} < \frac{D[h]}{D[\tilde{h}]} \leq \frac{\lambda_1 + 1}{\lambda_1 - 1}$$

M. Schiffer

where λ_1 is the lowest positive nontrivial Fredholm eigenvalue. The values $\lambda_\nu +1/ \lambda_\nu -1$ are stationary values of this functional. Because of the complete symmetry of D and $\tilde{D}$, the same values must be the stationary values of the reciprocal ratio or, in other words, the values $\lambda_\nu -1/ \lambda_\nu +1$ must likewise be stationary values for the original ratio. This is, indeed, the case since with each Fredholm eigenvalue $\lambda_\nu \neq -1$ also the Fredholm eigenvalue $-\lambda_\nu$ will occur. This is best seen from equations (7) and (8). Let $g_\nu(z)$ and $\tilde{g}_\nu(z)$ be the conjugate harmonic functions of $h_\nu(z)$ and $\tilde{h}_\nu(z)$. Then (7) and (8) yield by use of the Cauchy-Riemann equations.

$$(7') \qquad \frac{\partial g_\nu(z_0)}{\partial n} = \frac{1+\lambda_\nu}{1-\lambda_\nu} \frac{\partial \tilde{g}_\nu(z_0)}{\partial n}$$

and

$$(8') \qquad g_\nu(z_0) = \tilde{g}_\nu(z_0)$$

if the additive constants in g_ν and $\tilde{g}_\nu$ are properly adjusted. Hence, if we define

$$(12) \qquad h_\nu^*(z) = \frac{1}{1+\lambda_\nu} g_\nu(z) , \qquad \tilde{h}_\nu^*(z) = \frac{1}{1-\lambda_\nu} \tilde{g}_\nu(z)$$

we find

$$(13) \qquad h_\nu^*(z) = \frac{1-\lambda_\nu}{1+\lambda_\nu} \tilde{h}_\nu^*(z) , \qquad \frac{\partial h_\nu^*}{\partial n} = \frac{\partial \tilde{h}_\nu^*}{\partial n}$$

which coincides with (7) and (8) if we replace λ_ν by $-\lambda_\nu$.

Since Fredholm eigenvalues occur always in pairs, we shall from now on understand by λ_ν the positive eigenvalue of the pair. Likewise, the $h_\nu(z)$ shall be the eigenfunctions belonging to the positive eigenvalues. Their conjugates will then automatically belong to $-\lambda_\nu$.

M. Schiffer

We introduce now the analytic functions

$$(14) \qquad v_\nu(z) = \frac{\partial}{\partial z} h_\nu(z) , \qquad \tilde{v}_\nu(z) = \frac{\partial}{\partial z} \tilde{h}_\nu(z)$$

where $\dfrac{\partial}{\partial z} = \dfrac{1}{2} \left(\dfrac{\partial}{\partial x} - i \dfrac{\partial}{\partial y} \right)$ is the well-known complex differentiator. A simple calculation leads to the following result :

$$(15) \qquad v_\nu(z) = \frac{\lambda_\nu}{\pi} \iint_D \frac{\overline{v_\nu(\zeta)}}{(\zeta - z)^2} \, d\tau , \qquad \tilde{v}_\nu(z) = - \frac{\lambda_\nu}{\pi} \iint_{\tilde{D}} \frac{\overline{\tilde{v}_\nu(\zeta)}}{(\zeta - z)^2} \, d\tau$$

Clearly, the trivial eigenvalue λ_o has to be discarded since its eigenfunction $h_o(z)$ is constant and is annihilated by differentiation. We understand the improper integrals (15) in the sense of a Cauchy principal value : we exclude first in the integration a circle of radius ε around z and then take the limit value of the integral as $\varepsilon \longrightarrow 0$. With this understanding (15) represents integral equations for the eigenfunctions $v_\nu(z)$, and we may restrict ourselves to the case of positive eigenvalues $\lambda_\nu > 0$.

It is also easily verified that

$$(16) \qquad \tilde{v}_\nu(z) = \frac{\lambda_\nu}{\pi(1 + \lambda_\nu)} \iint_D \frac{\overline{v_\nu(\zeta)}}{(\zeta - z)^2} \, d\tau \qquad \text{for } z \in \tilde{D}$$

$$v_\nu(z) = \frac{\lambda_\nu}{\pi(1 - \lambda_\nu)} \iint_{\tilde{D}} \frac{\overline{\tilde{v}_\nu(\zeta)}}{(\zeta - z)^2} \, d\tau \qquad \text{for } z \in D$$

These integrals are proper and display clearly the analytic character of the eigenfunctions.

Another great advantage of introducing the harmonic eigenfunctions $h_\nu(z)$ and the analytic eigenfunctions $v_\nu(z)$ comes from the important orthogonality relations which these functions satisfy. Observe that the kernel $k(z, \zeta)$ is not symmetric in its arguments and hence that the eigenfunctions $\varphi_\nu(\zeta)$ defined on C are, in general, not orthogonal. On the other hand, let $h_\nu(z)$ and $h_\mu(z)$ belong to different eigenvalues : $\lambda_\nu \neq \lambda_\mu$. We compute the Dirichlet integral

$$(17) \qquad D\left[h_\nu, h_\mu\right] = \iint\limits_D \nabla h_\nu \cdot \nabla h_\mu \, d\tau = - \int\limits_C h_\nu \frac{\partial h_\mu}{\partial n} \, ds$$

Using the boundary relations (7) and (8), we can write

$$(18) \qquad D\left[h_\nu, h_\mu\right] = \frac{\lambda_\nu + 1}{\lambda_\nu - 1} \int\limits_C \tilde{h}_\nu \frac{\partial \tilde{h}_\mu}{\partial n} \, ds = \frac{\lambda_\nu + 1}{\lambda_\nu - 1} D\left[\tilde{h}_\nu, \tilde{h}_\mu\right]$$

But both Dirichlet integrals are symmetric in ν and μ while we know by assumption that $\dfrac{\lambda_\nu + 1}{\lambda_\nu - 1} \neq \dfrac{\lambda_\mu + 1}{\lambda_\mu - 1}$. Hence, we proved

$$(19) \qquad D\left[h_\nu, h_\mu\right] = D\left[\tilde{h}_\nu, \tilde{h}_\mu\right] = 0 \qquad \text{for} \quad \lambda_\nu \neq \lambda_\mu$$

We may then assume without loss of generality that

$$(19') \qquad D\left[h_\nu, h_\mu\right] = D\left[\tilde{h}_\nu, \tilde{h}_\mu\right] = 0 \qquad \text{for} \quad \nu \neq \mu$$

From definition (14) we conclude next the orthogonality relations for the analytic eigenfunctions

$$(20) \qquad \iint\limits_D v_\nu(z) \overline{v_\mu(z)} \, d\tau = \iint\limits_D \tilde{v}_\nu(z) \overline{\tilde{v}_\mu(z)} \, d\tau = 0 \qquad \text{for} \quad \nu \neq \mu$$

We should like to normalize the eigenfunctions $h_\nu(z)$ in the Dirichlet norm. Clearly, in view of (18) we shall have to multiply $h_\nu(z)$ and $\tilde{h}_\nu(z)$

M. Schiffer

with different normalizing factors. Correspondingly, the $v_\nu(z)$ and $\tilde{v}_\nu(z)$ go over into analytic functions $w_\nu(z)$ and $\tilde{w}_\nu(z)$ with the following properties :

$$(21) \qquad w_\nu(z) = \frac{\lambda_\nu}{\pi} \iint_D \frac{\overline{w_\nu(\zeta)}}{(\zeta-z)^2} \, d\tau \; , \quad \tilde{w}_\nu(z) = \frac{\lambda_\nu}{\pi\sqrt{\lambda_\nu^2-1}} \iint_D \frac{\overline{w_\nu(\zeta)}}{(\zeta-z)^2} \, d\tau$$

$$\tilde{w}_\nu(z) = \frac{\lambda_\nu}{\pi} \iint_{\tilde{D}} \frac{\overline{\tilde{w}_\nu(\zeta)}}{(\zeta-z)^2} \, d\tau \; , \quad w_\nu(z) = \frac{\lambda_\nu}{\pi\sqrt{\lambda_\nu^2-1}} \iint_D \frac{\overline{\tilde{w}_\nu(\zeta)}}{(\zeta-z)^2} \, d\tau$$

and the orthonormalization

$$(22) \qquad \iint_D w_\nu \, \overline{w_\mu} \, d\tau = \delta_{\nu\mu} \qquad \iint_{\tilde{D}} \tilde{w}_\nu \, \overline{\tilde{w}_\mu} \, d\tau = \delta_{\nu\mu}$$

3. <u>Fredholm Eigenvalues and Hilbert Transforms</u>

We can now connect our results with some important general theorems of analysis. Let $f(z)$ be an arbitrary complex-valued function defined in the entire complex z-plane and of class $\mathcal{L}^2$.

Define its so-called Hilbert transform

$$(23) \qquad F(z) = \frac{1}{\pi} \iint \frac{\overline{f(\zeta)}}{(\zeta-z)^2} \, d\tau$$

This will be a new function with the same properties as $f(z)$ and with the same norm

$$(24) \qquad \|F\|^2 = \iint |F|^2 \, d\tau = \iint |f|^2 \, d\tau = \|f\|^2$$

M. Schiffer

The Hilbert transformation is an involution, that is, the Hilbert transform
of $F(z)$ is again $f(z)$. Finally, wherever $f(z)$ is analytic, its transform
$F(z)$ will be analytic too.

Consider the function $f(z)$ defined as $w_\nu(z)$ in D and as 0 in $\tilde{D}$.
Clearly, its Hilbert transform in $\dfrac{1}{\lambda_\nu}\, w_\nu(z)$ in D and $\dfrac{\sqrt{\lambda_\nu^2-1}}{\lambda_\nu}\, \tilde{w}_\nu(z)$ in
$\tilde{D}$. We may interpret the eigenfunctions $w_\nu(z)$ as the eigenfunctions of the
Hilbert transformation restricted to D and to the class of analytic functions
in D.

Let now $g(z)$ be a real-valued function in D which vanishes on the
boundary C of D and whose complex derivative $\dfrac{\partial g}{\partial z}$ is in D of the class
$\mathcal{L}^2$. It is easily verified that for $z \in D$

$$(25) \qquad \frac{1}{\pi} \iint\limits_{D} \left(\overline{\frac{\partial g(\zeta)}{\partial \zeta}} \right) \frac{1}{(\zeta-z)^2}\, d\tau = \frac{\partial g(z)}{\partial z}$$

such that all such functions $\dfrac{\partial g}{\partial z}$ are likewise eigenfunctions of the Hilbert
transformation with the eigenvalue 1. However, if $v(z)$ is an arbitrary ana-
lytic function in D with a finite norm, we have

$$(26) \qquad \iint\limits_{D} v(z)\, \overline{\left(\frac{\partial g}{\partial z}\right)}\, d\tau = 0$$

Hence, the linear space of all analytic functions with finite norm is orthogonal to these eigenfunctions $\dfrac{\partial g}{\partial z}$.

The linear space of all complex valued functions in D of class $\mathcal{L}^2$
can be split into the two complementary subspaces consisting of the $\dfrac{\partial g}{\partial z}$ and
of analytic functions. It is evident that the nontrivial part of the theory of
Hilbert transforms belongs to the subspace of analytic functions and not to
the trivial orthogonal complement where it reduces to the identity transfor-

M. Schiffer

mation. The theory of the Hilbert transform in the subspace of analytic functions was developed by Bergman and Schiffer $[3, 5]$. The general theory for the $\mathcal{L}^2$-space was first indicated by Beurling $[2, 4]$.

We shall see that the Hilbert transformation in the analytic subspace can be reduced to an integral transformation with a completely continuous kernel.

4. The Green's Function and its Analytical Kernels

Let $g(z, \zeta)$ be the harmonic Green's function of D. That is, $g(z, \zeta)$ is harmonic in both arguments for $z \neq \zeta$, vanishes if either argument point lies on the boundary C of D and behaves such that $g(z, \zeta) + \log|z - \zeta|$ is regular harmonic as $z \to \zeta$. It is well known that $g(z, \zeta)$ is symmetric in both arguments. We define now the two kernels $[3, 15]$

$$(27) \qquad K(z, \bar{\zeta}) = -\frac{2}{\pi} \frac{\partial^2 g}{\partial z \, \partial \bar{\zeta}} , \qquad L(z, \zeta) = -\frac{2}{\pi} \frac{\partial^2 g}{\partial z \, \partial \zeta}$$

K is hermitian in the two variables, analytic in z and antianalytic in ζ. It is regular even for $z = \zeta$ since the differentiation process which defines K annihilates the singularity of the Green's function. On the other hand, $L(z, \zeta)$ is analytic and symmetric in its variables, but it has a double pole at $z = \zeta$ and can be written as

$$(28) \qquad L(z, \zeta) = \frac{1}{\pi (z - \zeta)^2} - \ell(z, \zeta)$$

Here $\ell(z, \zeta)$ is regular analytic and symmetric in both variables. It is even continuous in the closure of D. If C is an analytic curve, it is even analytic in $D + C$.

M. Schiffer

From the boundary behavior of the Green's function a simple integration by parts leads to the following identities valid for every analytic function $\varphi(z)$ with finite norm over D :

$$(29) \qquad \iint_D K(z, \overline{\zeta}) \, \varphi(\zeta) \, d\tau = \varphi(z), \qquad \iint_D L(z, \zeta) \, \overline{\varphi(\zeta)} \, d\tau = 0$$

This shows that $K(z, \overline{\zeta})$ is the Bergman kernel function which reproduces every analytic function with finite norm; $L(z, \zeta)$ annihilates the same function class under the integration considered. We may rephrase the second identity as follows :

$$(30) \qquad \frac{1}{\pi} \iint_D \frac{\overline{\varphi(\zeta)}}{(\zeta - z)^2} \, d\tau = \iint_D \ell(z, \zeta) \, \overline{\varphi(\zeta)} \, d\tau$$

On the left side stands the improper integral which defines according to (23) the Hilbert transform of $\varphi(z)$. On the right we have an integral transformation which is completely continuous and coincides with the Hilbert transform on the subspace of all analytic functions in D with finite norm. This new definition of the Hilbert transform on the subspace is, of course, of very great convenience.

Let us consider an arbitrary complete orthonormal system $w_\nu(z)$ in the subspace of analytic functions in D. The Bergman kernel $K(z, \overline{\zeta})$ can be developed into a Fourier series in the system, and we have by virtue of (29)

$$(31) \qquad K(z, \overline{\zeta}) = \sum_{\nu=1}^{\infty} w_\nu(z) \, \overline{w_\nu(\zeta)}$$

This was, indeed, Bergman's original definition of his kernel function. It is easy to see that the Fredholm eigenfunctions $w_\nu(z)$ defined by (21) and (22)

M. Schiffer

form a complete set and may be used in the representation (31).

Moreover, we have for this particular choice of the orthonormal system in view of (21) and (30)

$$(32) \qquad w_\nu(z) = \lambda_\nu \iint_D \ell(z, \zeta) \overline{w_\nu(\zeta)} \, d\tau$$

We can express $\ell(z, \zeta)$ for z fixed as a Fourier series in the $w_\nu(\zeta)$ and (32) yields us the Fourier coefficients. We have then

$$(33) \qquad \ell(z, \zeta) = \sum_{\nu=1}^{\infty} \frac{w_\nu(z) w_\nu(\zeta)}{\lambda_\nu}$$

We are led next to an important and beautiful identity for the ℓ-kernel by using the concept of the Hilbert transform. Let $f(z)$ be analytic in D and of finite norm. We may conceive it as a complex valued function of class $\mathscr{L}^2$ in the entire plane if we define it as identically zero in $\tilde{D}$. Its Hilbert transform $F(z)$ can be written as follows :

$$(34) \qquad F(z) = \iint_D \ell(z, \zeta) \overline{f(\zeta)} \, d\tau \qquad\qquad \text{if } z \in D$$

$$F(z) = \frac{1}{\pi} \iint_D \frac{\overline{f(\zeta)}}{(\zeta - z)^2} \, d\tau \qquad\qquad \text{if } z \in \tilde{D}$$

The identity of norms (24) for Hilbert transforms yield thus

$$(35) \qquad \iint_D |f(z)|^2 \, d\tau = \iint_D \iint_D K(\zeta, \overline{\eta}) \overline{f(\zeta)} f(\eta) \, d\tau_\eta \, d\tau_\zeta$$

$$= \iint_D \iint_D \iint_D \ell(z, \zeta) \overline{\ell(z, \eta)} f(\eta) \overline{f(\zeta)} \, d\tau_z \, d\tau_\zeta \, d\tau_\eta$$

$$+ \iint_{\tilde{D}} \iint_D \iint_D \frac{1}{\pi^2} \frac{1}{(z-\zeta)^2} \frac{1}{\overline{(z-\eta)}^2} f(\eta) \overline{f(\zeta)} \, d\tau_z \, d\tau_\zeta \, d\tau_\eta$$

M. Schiffer

A standard argument leads, therefore, to the identity

$$(36) \qquad \iint_D \ell(z, \zeta) \, \overline{\ell(z, \zeta)} \, d\tau + \frac{1}{\pi^2} \iint_{\widetilde{D}} \frac{d\tau}{(z-\zeta)^2 \overline{(z-\eta)}^2} \, d\sigma = K(\zeta, \overline{\eta})$$

Observe that the second left-hand integral is regular analytic for $\zeta \in D$ and regular anti-analytic for $\eta \in D$. It can be computed by integrations and is, therefore, more elementary than the kernels K and ℓ which depend on the Green's function of the domain, that is, on the solution of a boundary value problem for harmonic functions. We shall call the expression

$$(37) \qquad \Gamma(\zeta, \overline{\eta}) = \frac{1}{\pi^2} \iint_{\widetilde{D}} \frac{d\tau}{(z-\zeta)^2 \overline{(z-\eta)}^2}$$

a geometric term in contradistinction to the more trascendental kernels K and ℓ. Clearly, Γ is hermitian and a positive definite kernel. If we insert into (36) the Fourier developments (31) and (33), we find the Fourier development for the geometric kernel

$$(38) \qquad \Gamma(\zeta, \overline{\eta}) = \sum_{\nu=1}^{\infty} \left(1 - \frac{1}{\lambda_\nu^2}\right) w_\nu(\zeta) \, \overline{w_\nu(\eta)}$$

This representation may serve as a basis for calculating the kernels ℓ and K. The basic idea is as follows. All numbers $\xi_\nu = \left(1 - \frac{1}{\lambda_\nu^2}\right)$ lie in the interval $0 < \left(1 - \frac{1}{\lambda_1^2}\right) \leq \xi \leq 1$. In this interval $\frac{1}{\xi} = \sum_{n=0}^{\infty} (1-\xi)^n$ converges absolutely and uniformly. Thus,

$$(39) \qquad 1 = \xi \sum_{n=0}^{\infty} (1-\xi)^n$$

M. Schiffer

can be approximated uniformly and arbitrarily by polynomials $P_N(\zeta) =$

$$\sum_{\nu=0}^{N} a_{N\nu}\, \zeta^{\nu}.$$ If $\Gamma^{(\nu)}(\zeta,\overline{\eta})$ denotes the νth iterate of the kernel

$\Gamma(\zeta,\overline{\eta}) = \Gamma^{(1)}(\zeta,\overline{\eta})$ it is clear that

$$\text{(40)} \qquad \sum_{\nu=0}^{N} a_{N\nu}\Gamma^{(\nu)}(\zeta,\overline{\eta}) \underset{N}{\longrightarrow} K(\zeta,\overline{\eta})$$

Thus, the kernel K and, as is easily seen, the kernel ℓ can be approximated arbitrarily in terms of iterates of the elementary geometric kernel $\Gamma(\zeta,\overline{\eta})$ [3, 15] .

5. Fredholm Eigenvalues and Univalent Functions

Let us suppose that $f(z)$ is analytic and univalent in the unit circle and maps $|z| < 1$ onto a domain D in the w-plane. Because of the conformal invariance of the Green's function we have for the L-kernel of D the identity

$$\text{(41)} \qquad L_D(w,\omega)f'(z)f'(\zeta) = L(z,\zeta)$$

with $\omega = f(\zeta)$ and $L(z,\zeta)$ being the L-kernel of the unit circle. Since the Green's function of a circle is well known, we have

$$\text{(42)} \qquad L(z,\zeta) = \frac{1}{\pi(z-\zeta)^2}$$

and hence

M. Schiffer

$$(43) \quad \ell_D(w, \omega) f'(z) f'(\zeta) = \frac{1}{\pi} \left[\frac{f'(z) f'(\zeta)}{(f(z)-f(\zeta))^2} - \frac{1}{(z-\zeta)^2} \right]$$

$$= \frac{1}{\pi} \frac{\partial^2}{\partial z \, \partial \zeta} \log \frac{f(z)-f(\zeta)}{z-\zeta}$$

Now, the function

$$(44) \quad \log \frac{f(z)-f(\zeta)}{z-\zeta} = \sum_{\mu, \nu = 0}^{\infty} c_{\mu\nu} \, z^\mu \zeta^\nu$$

plays a central role in the theory of univalent functions. Indeed, a necessary and sufficient condition that $f(z)$ be regular and univalent in $|z| < 1$ is the regularity of this function in two complex variables in the product domain $|z| < 1, |\zeta| < 1$. This important formulation of univalency was observed a long time ago, but Grunsky [7] was the first to draw important conclusions from it. Using the preceding relations between the ℓ- and K-kernel (in particular (36)), we can derive the Grunsky inequalities

$$(45) \quad \left| \sum c_{\mu\nu} \, x_\mu \, x_\nu \right| \le \sum \nu |x_\nu|^2$$

for arbitrary complex vectors x_ν as the necessary and sufficient condition for the univalence of $f(z)$. From these conditions many elementary estimates for the coefficients of univalent functions may be obtained; the most startling one is $|a_4| \le 4$ proved by Charzynski and Schiffer [5].

M. Schiffer

We can bring every quadratic form $Q(x,x)$ with symmetric matrix Q into the normal form

$$(46) \qquad Q(x,x) = \sum \ell_\nu \, y_\nu^2 \, , \qquad\qquad \ell_\nu > 0$$

where the y_ν are obtained from the x_ν by a unitary transformation. In other words, every simmetric matrix Q can be brought into the Schur normal form $[16]$ $Q = U^T \triangle U$ where U is a unitary matrix, U^T is its transposed and $\triangle$ is a diagonal matrix with positive elements ℓ_ν . There arises then the question : Given a univalent function $f(z)$ in $|z| < 1$ and its corresponding symmetric matrix $c_{\mu\nu}$ defined by (44), what are the positive numbers ℓ_ν which appear in its Schur normal form? The answer is : The ℓ_ν are simple expressions in the Fredholm eigenvalues λ_ν of the image domain D. This surprising relation can be read off from (33), (43) and (44). It shows clearly the great significance of the Fredholm eigenvalues for the general theory of analytic functions and conformal mapping.

6. The Variation of Fredholm Eigenvalues

The most powerful tool in the study of Fredholm eigenvalues are the variational formulas which show how the λ_ν behave as functionals of the curve C and how they vary with continuously changing boundary. To find the functional derivatives of the λ_ν with respect to C we transform the integral equation (21) for $w_\nu(z)$ by partial integration into

M. Schiffer

$$(47) \qquad w_\nu(z) = \frac{\lambda_\nu}{2\pi i} \oint_C \frac{\overline{(w_\nu(\zeta)d(\zeta))}}{\zeta - z}$$

We select an arbitrary but fixed point $z_0 \in \tilde{D}$ and consider the variation

$$(48) \qquad z^* = z + \frac{\varepsilon}{z - z_0}$$

The mapping $z^*(z)$ is, for small enough $|\varepsilon|$, a regular analytic function of z in $D + C$ and maps D into a new domain D^* with boundary C^*, eigenvalues λ_ν^* and eigenfunctions $w_\nu^*(z)$. If we can give an asymptotic formula for λ_ν^*, we shall have achieved our aim since the most general variation of C may be obtained by superposition of elementary variations (48).

We write down the corresponding integral equation (47) for λ_ν^*. But putting $z^* = z^*(z)$, $\zeta^* = \zeta^*(\zeta)$, we may refer the integration back to the original curve C and obtain an integral equation for the fixed domain D whose kernel now depends on z_0 and ε. We find

$$(49) \qquad w^*(z^*(z)) = \frac{\lambda_\nu^*}{2\pi i} \oint_C \frac{\overline{\left(w_\nu^*(\zeta^*(\zeta))(1 - \frac{\varepsilon}{(\zeta - z_0)^2})d\zeta\right)}}{(\zeta - z)\left(1 - \frac{\varepsilon}{(z-z_0)(\zeta - z_0)}\right)}$$

Let us define the new unknown function in D

M. Schiffer

$$(50) \qquad m_\nu(z) = w_\nu^*(z^*(z)) \left(1 - \frac{\varepsilon}{(z-z_0)^2}\right)$$

and bring (49) into the much simpler form

$$(51) \qquad m_\nu(z) = \frac{\lambda_\nu^*}{2\pi i} \oint_C \overline{(m_\nu(\zeta)d\zeta)} \left[\frac{1}{\zeta-z} + \frac{\varepsilon}{(z-z_0)\left[(z-z_0)(\zeta-z_0)-\varepsilon\right]}\right]$$

We thus have for $m_\nu(z)$ an integral equation whose kernel depends analytically on ε . We may also return to the domain integral form

$$(52) \qquad m_\nu(\bar{z}) = \frac{\lambda_\nu^*}{\pi} \iint_D \overline{m_\nu(\zeta)} \left[\frac{1}{(\zeta-z)^2} + \frac{\varepsilon}{(\zeta-z_0)^2(z-z_0)^2} + 0(\varepsilon^2)\right] d\tau$$

We may replace the first term $\frac{1}{\pi}(\zeta-z)^{-2}$ in the kernel of (52) by $\ell(z,\zeta)$. Thus, this integral equation falls well into the pattern of the Rellich theory $\begin{bmatrix}8\end{bmatrix}$ for eigenvalues of variable kernels. We may assert that

$$(53) \qquad \lambda_\nu^* = \lambda_\nu + |\varepsilon|\, \ell_\nu + \ldots$$

admits a power series development in ε . For the sake of simplicity, let us assume that λ_ν is nondegenerate, i.e., it has only one eigenfunction $w_\nu(z)$. In this case we also have a series development

M. Schiffer

$$(54) \qquad m_{\nu}(z) = w_{\nu}(z) + |\varepsilon|\, \omega_{\nu}(z) + \ldots$$

We now multiply equation (52) by $\overline{w_{\nu}(z)}$ and integrate over D. We use integral equation (21) for $w_{\nu}(z)$ and observe the asymptotics in ε as given in (53) and (54). Thus,

$$(55) \qquad \iint\limits_{D} m_{\nu}(z)\,\overline{w_{\nu}(z)}\, d\tau = \frac{\lambda_{\nu}^{*}}{\lambda_{\nu}} \iint\limits_{D} \overline{m_{\nu}(\zeta)} w_{\nu}(\zeta)\, d\tau$$

$$+\, \varepsilon\, \frac{\lambda_{\nu}}{\pi} \iint\limits_{D} \frac{\overline{w_{\nu}(\zeta)}}{(\zeta - z_0)^2}\, d\tau \cdot \iint\limits_{D} \frac{\overline{w_{\nu}(z)}}{(z - z_0)^2}\, d\tau + 0(\varepsilon^2)$$

By our assumption $z_0 \in \tilde{D}$, and hence using again the corresponding identity (21), we find

$$(56) \qquad \lambda_{\nu} \iint\limits_{D} m_{\nu}(z)\overline{w_{\nu}(z)}\, d\tau = \lambda_{\nu}^{*} \iint\limits_{D} \overline{m_{\nu}(\zeta)} w_{\nu}(\zeta)\, d\tau +$$

$$+\, \pi\varepsilon\, \tilde{w}_{\nu}(z_0)^2 (\lambda_{\nu}^2 - 1) + 0(\varepsilon^2)$$

Take the real parts in (56), use (54) and the normalization of $w_{\nu}(z)$ to find

$$(57) \qquad \lambda_{\nu}^{*} - \lambda_{\nu} = -\,\mathrm{Re}\,\left\{ \pi\,\varepsilon\,(\lambda_{\nu}^2 - 1)\tilde{w}_{\nu}(z_0)^2 \right\} + 0(\varepsilon^2)$$

This is the desired variational formula for λ_{ν}. A similar formula may be given if λ_{ν} is a degenerate eigenvalue. Had we taken a

M. Schiffer

variation (48) with $z_0 \in D$, we might have reasoned in the same way by starting with the integral equation for $\tilde{w}_\nu(z)$ over $\tilde{D}$. We would have found the analogous formula

$$(58) \qquad \delta \lambda_\nu = - \mathrm{Re} \quad \{ \pi \varepsilon (\lambda_\nu^2 - 1) w_\nu (z_0)^2 \} + 0(\varepsilon^2)$$

for the variation of a nondegenerate eigenvalue.

To illustrate the power of these variational formulas, we quote one extremum problem which has been solved by using them. Let $f(z)$ be univalent and regular in the circular ring $r \le |z| \le R$ with $r < 1 < R$. Let C be the image of $|z| = 1$ by this map. We call C a uniformly analytic curve with the modulus (r, R). The concept of uniform analyticity is an obvious sharpening of the usual assumption of analyticity of a curve. We have the theorem :

The lowest nontrivial eigenvalue λ_1 of a uniformly analytic curve with the modulus (r, R) satisfies the inequality

$$(59) \qquad \lambda_1 \ge \frac{r^2 + R^2}{1 + r^2 R^2}$$

This estimate is the best possible. It shows the importance of the concept of uniform analyticity in the numerical procedures of conformal mapping.

7. Fredholm Eigenvalues for Multiply-Connected Domains

It is easy to extend the preceding considerations to the case that $\tilde{D}$ is a multiply-connected domain which contains the point at

M. Schiffer

infinity and is bounded by $N > 1$ closed curves C_ν , which shall again be three times continuously differentiable. Its complement D will then consist of N disjoint simply-connected finite domains D_ν . Let $C = \sum C_\nu$ denote the common boundary of $\tilde{D}$ and D . We may then discuss the Fredholm eigenvalue problem (5) with respect to the curve system C .

We may extend again the eigenfunctions of this problem which are defined only on C into harmonic functions in $\tilde{D}$ and in D . As before, we can interpret the eigenvalues λ_ν also as eigenvalues of integral equations for analytic functions of the Hilbert transform type. We can define a function $w_\nu(z)$ defined and analytic in each component D_ν of D such that

$$(60) \qquad w_\nu(z) = \frac{\lambda_\nu}{\pi} \iint\limits_{D} \frac{\overline{w_\nu(\zeta)}}{(\zeta - z)^2} \, d\tau , \qquad z \in D$$

and one single analytic function $\tilde{w}_\nu(z)$ in the connected domain $\tilde{D}$ such that

$$(61) \qquad \tilde{w}_\nu(z) = \frac{\lambda_\nu}{\pi} \iint\limits_{\tilde{D}} \frac{\overline{\tilde{w}_\nu(\zeta)}}{(\zeta - z)^2} \, d\tau , \qquad z \in \tilde{D}$$

These functions are related to the Fredholm eigenfunctions $\varphi_\nu(z)$ of (5) in the same way as in the case of simple connectivity. There is, however, one important difference. The eigenvalue $\lambda = 1$ occurs in (5) in $(N-1)^{st}$ order degeneracy. The integral equation (60) does not possess this eigenvalue at all while $\lambda = 1$ is an eigenvalue of (61) of order $N - 1$. The corresponding eigenfunctions are the derivatives

M. Schiffer

of the harmonic measures of the multiply-connected domain $\tilde{D}$. Since $\lambda = 1$ leads to simple and well-known eigenfunctions, we still shall call it the trivial eigenvalue and assume in all subsequent discussions $\lambda_\nu > 1$.

It is easily seen that for $\lambda_\nu > 1$ still

$$(62) \qquad \tilde{w}_\nu(z) = \frac{\lambda_\nu}{\pi\sqrt{\lambda_\nu^2 - 1}} \iint\limits_{D} \frac{\overline{w_\nu(\zeta)}}{(\zeta - z)^2}\, d\tau, \qquad z \in \tilde{D}$$

$$w(z) = \frac{\lambda_\nu}{\pi\sqrt{\lambda_\nu^2 - 1}} \iint\limits_{\tilde{D}} \frac{\overline{\tilde{w}_\nu(\zeta)}}{(\zeta - z)^2}\, d\tau, \qquad z \in D$$

and that the $w_\nu(z)$ and $\tilde{w}_\nu(z)$ form orthonormal systems in their respective domains.

Finally, we can extend the entire theory of the Hilbert transform by means of $\ell(z, \zeta)$ to the case of the connected region $\tilde{D}$. However, if we wish to do the same thing for the set of domains D_ν, we first have to give a proper definition of $\ell(z, \zeta)$. We start with defining a Green's function $g(z, \zeta)$ for the disconnected region D, namely

$$(63) \qquad g(z, \zeta) = \begin{cases} g_\nu(z; \zeta) & \text{if } z, \zeta \text{ lie in same } D_\nu \\ \\ 0 & \text{if } z, \zeta \text{ lie in different } D_\nu \end{cases}$$

M. Schiffer

Here $g_\nu(z, \zeta)$ is the ordinary Green's function of the simply - connec-ted domain D_ν. We define next $L(z, \zeta)$ from $g(z, \zeta)$ by means of (27) and $\ell(z, \zeta)$ by means of (28). Thus

$$(64) \qquad \ell(z, \zeta) = \begin{cases} \ell_\nu(z, \zeta) & \text{if } z, \zeta \text{ lie in same } D_\nu \\[2em] \dfrac{1}{\pi(z-\zeta)^2} & \text{if } z, \zeta \text{ lie in different } D_\nu \end{cases}$$

$\ell_\nu(z, \zeta)$ is the ℓ-kernel of the simply-connected domain D_ν. With this definition, (30) remains obviously valid, and the Fourier re-presentation (33) of $\ell(z, \zeta)$ in terms of the analytic Fredholm eigen-functions is preserved.

The variational formulas of the preceding section can be carried over without change since we did not use anywhere in our calculations that C consists of one single curve.

8. Fredholm Determinants and Conformal Mapping

Having enumerated many definitions and identities, we shall now show their usefulness and interest by particular applications. An impor-tant concept in integral equation theory is the Fredholm determinant $\begin{bmatrix} 12, 14 \end{bmatrix}$

$$(65) \qquad D(\lambda) = \prod_{\nu=1}^{\infty} (1 - \frac{\lambda^2}{\lambda_\nu^2})$$

where the product is extended over all nontrivial eigenvalues λ_ν. We observe that the eigenvalues $+\lambda_\nu$ and $-\lambda_\nu$ occur always in pairs in

M. Schiffer

our problem; this accounts for the quadratic factors.

Consider $D(\lambda)$ for fixed λ as a functional of C and ask for its variation if C is varied by the standard variation (48) with $z_0 \in D$. By virtue of (58) we find

$$(66) \qquad \delta \log D(\lambda) = - \operatorname{Re} \left\{ 2 \pi \varepsilon \lambda^2 \sum_{\nu=1}^{\infty} \frac{\lambda_\nu^2 - 1}{\lambda_\nu^2 - \lambda} \frac{w_\nu(z_0)^2}{\lambda_\nu} \right\}$$

This formula can also be justified if some λ_ν are degenerate eigenvalues. The result simplifies considerably in the case $\lambda = 1$. Indeed, in view of (33) we can write

$$(67) \qquad \delta \log D(1) = - \operatorname{Re} \left\{ 2 \pi \varepsilon \ell(z_0, z_0) \right\}$$

Thus the important function $\ell(z, z)$ has been identified as the funtional derivative of the logarithm of the Fredholm determinant.

A surprising result occurs in the case of a multiple connectivity $N > 1$. We can speak of the eigenvalues of the curve system C and their Fredholm determinant; we may also consider the eigenvalues $\lambda_\nu^{(k)}$ of the single curve C_k and their Fredholm determinant $D^{(k)}(1)$. If $z_0 \in D_k$ we have by (64) the identity

$$(68) \qquad \delta \log D(1) = \delta \log D^{(k)}(1)$$

That is, under a standard variation (48) which is regular analytic outside of D_k, the ratio $D^k(1) / D(1)$ has zero variation. By reasoning typical for variational theory, we can then extend this result to arbitrary finite conformal maps in the exterior of D_k.

M. Schiffer

Theorem. Let D be a set of disjoint finite simply-connected domains D_ℓ with the boundary curve system C. Let $D(1)$ and $D^{(\ell)}(1)$ be the Fredholm determinants of the curve system C and the single curve C_ℓ, respectively. Let $w = f(z)$ be a conformal mapping in the exterior of D_ℓ. It will carry the curve system C into a curve system Γ; let $\Delta(1)$ and $\Delta^{(\ell)}(1)$ be the corresponding Fredholm determinants of Γ and Γ_ℓ (the image of C_ℓ). Then

$$(69) \qquad \frac{\Delta^{(\ell)}(1)}{\Delta(1)} = \frac{D^{(\ell)}(1)}{D(1)}$$

It seems difficult to prove the conformal invariance of this ratio in a nonvariational manner.

We recall the fact that if C is a circle, all its eigenvalues are infinite and that the Fredholm determinant of each circle has the constant value 1. In every other case the definition (65) clearly indicates that $D(1) < 1$. Hence, suppose that we start with an arbitrary curve set C and map the exterior of C_ℓ conformally onto the exterior of a circle. By (69) we can assert for the Fredholm determinant $\Delta(1)$ of the new curve system

$$(70) \qquad \Delta(1) = \frac{D(1)}{D^{(\ell)}(1)} \geq D(1)$$

Equality in (76) holds only if C_ℓ already happened to be a circle.

This remark throws light on a well-known procedure to map a multiply-connected domain onto a circular domain. One starts with the curve C_1 and maps its exterior onto the exterior of a circle. Then one takes the image of C_2 and maps its exterior onto the exterior of a circle. One continues this procedure indefinitely taking care to run through the images of all starting curves in fixed order. The limit of this map transforms all initial curves C_ν into circles. We see that in this procedure the Fredholm determinant D(1) is steadily increased. One can base on this observation a convergence proof for this method of iteration. We also draw the following conclusion :

Theorem. Among all conformally equivalent domains the circular domain has the largest Fredholm determinant D(1).

This theorem was originally proved by variational methods [12] . The present derivation explains more clearly its significance.

9. **Conclusion.** The close relation between the Fredholm eigenvalue problem and the theory of analytic functions of one complex variable has been evident throughout the whole exposition. Hence, it will be expected that the potential theory in more than two dimensions will lead to Fredholm eigenvalues with a less elegant and elastic theory . However, many results can be preserved even in this transition . However, one very significant result shows the great difference in the nature of the eigenvalues for different dimensions.

Theorem. Let D be a domain in space and let λ_1 be its lowest positive nontrivial Fredholm eigenvalue. Then

$$\lambda_1 \leq 3 \tag{71}$$

Equality holds only in the case that D is a sphere [13] .

M. Schiffer

Thus, the Liouville - Neumann series development, which solves the boundary value problem in three-dimensional potential theory, will never converge better than a geometric series with ratio $\frac{1}{3}$.

Another significant difference comes from the fact that the concept of conjugate harmonic functions fails in more than two dimensions. Hence, we cannot assert that with each Fredholm eigenvalue λ_ν also its negative $-\lambda_\nu$ will occur as an eigenvalue.

The study of Fredholm eigenvalues in more than two dimensions is thus still an open and promising field of research.

BIBLIOGRAPHY

[1]   L.V. Ahlfors, Remarks on the Neumann-Poincaré integral equation, Pacific J. Math. $\underline{2}$ (1952), 271-280.

[2]   ——————, Conformality with respect to Riemann metrics, Ann. Acad. Fenn., Series A206 (1955).

[3]   S. Bergman and M. Schiffer, Kernel functions and conformal mapping, Compositio Math. $\underline{8}$ (1951), 205-249.

[4]   I.E. Block, Kernel functions and class L^2, Proc. Amer. Math. Soc. $\underline{4}$ (1953), 110-117.

[5]   S. Charzynski and M. Schiffer, A new proof of the Bieberbach conjecture for the fourth coefficient, Arch. Rational Mech. Anal. $\underline{5}$ (1960), 187-193.

[6]   S. Gershgorin, On conformal mapping of a simply-connected region onto a circle, Math. Sb. $\underline{40}$ (1933), 48-59.

[7]   H. Grunsky, Koeffizientenbedingungen für schlicht abbildende meromorphe Funktionen, Math. Z. $\underline{45}$ (1939), 29-61.

[8]   F. Rellich, Störungstheorie der Spektralzerlegung, I. Mitteilung, Math. Ann. $\underline{113}$ (1937), 600-619.

[9]   H.L. Royden, A modification of the Neumann-Poincaré method for multiply-connected regions, Pacific J. Math. $\underline{2}$ (1952), 385-394.

[10]   M. Schiffer, Applications of variational methods in the theory of conformal mapping, Proc. Symp. Appl. Math. $\underline{8}$ (1958), 93-113.

[11]   ——————, The Fredholm eigenvalues of plane domains, Pacific J. Math. $\underline{7}$ (1957), 1187-1225.

[12]   M. Schiffer, Fredholm eigenvalues of multiply-connected domains, Pacific J. Math. $\underline{9}$ (1959), 211-264.

[13]   ——————, Problèmes aux limites et fonctions propres de l'équation intégrale de Poincaré et de Fredholm, C. R. Paris $\underline{245}$ (1957), 18-21.

[14]   M. Schiffer and N. Hawley, Connections and conformal mapping, Acta Math. $\underline{107}$ (1962), 175-274.

[15]   M. Schiffer and D.C. Spencer, Functionals of finite Riemann surfaces, Princeton 1954.

[16]   J. Schur, Ein Satz über quadratische Formen mit komplexen Koeffizienten, Amer. J. Math. $\underline{67}$ (1945), 472-480.

[17]   S.E. Warschawski, On the effective determination of conformal maps, Contribution to the theory of Riemann surfaces, Princeton 1953.